Robert Etherridge

Records of the Australian Museum

Vol. III

Robert Etherridge

Records of the Australian Museum
Vol. III

ISBN/EAN: 9783743383265

Manufactured in Europe, USA, Canada, Australia, Japa

Cover: Foto ©berggeist007 / pixelio.de

Manufactured and distributed by brebook publishing software (www.brebook.com)

Robert Etherridge

Records of the Australian Museum

RECORDS

OF THE

AUSTRALIAN MUSEUM

EDITED BY THE CURATOR

Vol. III.

PRINTED BY ORDER OF THE TRUSTEES.

R. ETHERIDGE, JUNR,

Curator.

SYDNEY, 1897-1900.

No. 4.   Published 13th June, 1898.

No. 5.   Published 17th April, 1899.

CONTENTS.

No. 6. Published 11th December, 1899.

No. 7. Published 15th June, 1900.

Part 8. Published 1st December, 1900.

Title Page, Contents, and Index.

LIST OF THE CONTRIBUTORS.

With References to the Articles contributed by each.

LIST OF THE PLATES.

[NOTE.—For the convenience of those who prefer to bind the Plates with the text, rather than at the end of the volume, the pages which they should face are indicated in margin.]

LIST OF THE PLATES—*Continued.*

CORRECTIONS.

Page 16, line 3—for " *Columbia* " read " *Columba.* "

" 28, line 25 —for " *lenoscopelus* " read " *leucoscopelus.* "

" 29, line 5—for " *lenoscopelus* " read " *leucoscopelus.* "

" 29, line 16 for " *lenoscopelus* " read " *leucoscopelus.* "

" 67, line 34—for " meridianal " read " meridional."

" 188—*Megaderma gigas* has also been recorded from Central Australia.

" 195—Delete " *Epinephelus tauvina*, Forskal."

" 219, line 5—for " B. B. Woodward " read " B. H. Woodward."

Plate viii., Explanation—for "*Columnopora*" read "*Columnaria.*"

ERRATA.

Page 28, line 25, for " *lenoscopelus* " read " *leucoscopelus.* "

" 29, line 5, for „ read „

" 29, line 16, for „ read „

Plate XI., Explanation, line 2, for " *sublimais* " read " *sublimis.* "

On CIRCULAR and SPIRAL INCISED ORNAMENT on AUSTRALIAN ABORIGINAL IMPLEMENTS and WEAPONS.

By R. Etheridge, Junr., Curator.

(Plates i., ii.)

THE more or less rare occurrence of this form of sculpture on the implements and weapons of our Aborigines will probably render a notice of several instances interesting.

The late Mr. R. Brough Smyth remarked* many years ago that. – "Curved lines are rarely seen. Any attempt to represent a curve in all the specimens I have examined has been a failure.' Mr. Andrew Lang even made a more sweeping statement† when he wrote that the patterns used by the Australian Aborigines are such as can be produced without the aid of "spirals or curves or circles."

Of the incorrectness of this statement, no better example can be adduced than the circular incised figures seen on the "Bull-roarers" figured‡ by the late Mr. Edward Hardman, from the Kimberley District, N.W. Australia.

A very beautiful instance is represented in Pl. i., Fig. 1 and 2, all the more interesting because it is a stone implement, and the only one of its kind that has ever come under my notice. It consists of a flat pebble (in all probability) of indurated shale, long-oval in shape, and incised on both faces; five and six-eights inches long, and three and three-sixteenths wide, but is fractured at the lower end. On one aspect (Pl. i., Fig. 1) is a nearly central figure consisting of incised circles arranged spirally within one another. The figure is generally very slightly longer than wide, the greatest or longitudinal diameter being two and fifteen-sixteenths inches. On the right hand side there are thirteen incised grooves, and on the left twelve, the grooves becoming slightly wider towards the circumferential one. Immediately above, on the same face of the pebble, are two smaller figures, the incised grooves, three in number in each case, being however simply concentric within one another, and not spiral. That on the right is half-an-inch in diameter, and that on the left five-eighths.

* Smyth; Aborigines of Victoria, i., 1878, p. 283.
† Lang; Customs and Myth, p. 279.
‡ Proc. R. Irish Acad., i., 1888 (2), No. 1, t. 2, f. 4, t. 3.

Both above these circles, and below the large spiral, is a series of horizontal grooves, starting from the edge of the implement inwards, and not meeting in the middle line, but leaving a clear median space, that below the spiral being wider than the upper one. The upper incisions are seven on the right, and eight on the left, the lower six on either side.

The reverse of this implement (Pl. i., Fig. 2) is differently incised. In the place of the large almost central spiral is a rather roughly executed series of circles concentric within one another; the longest diameter is two and two-eighths inches. Surmounting this figure is a three-quarters circular representation in which the grooves are very much finer, fainter, and closer together, leaving a large unincised space, the free end almost touching the circumference of the central group of circles. If completed, the figure would also be circular in form. There are, I think, ten grooves. The surface of the implement below the central circles is transversely grooved in a manner precisely similar to that of the front face, the grooves on the left hand numbering thirteen, but those on the right are too indistinct for enumeration.

This peculiar implement has been profusely ruddled, so much so as to almost fill up some of the grooves. Furthermore, the broken base shows traces of adherent gum cement, which extends on the fractured end. I infer from this either that it has formed a portion of a mounted implement, or when in the possession of its sable owner was fractured and possibly repaired. The colour is fast, and does not soil the hand.

This interesting implement was presented some years ago by a Mr. Dunlop, and is said to come from North Queensland, but the precise locality is unknown.

As having a possible bearing on the use of this instrument, it is necessary to refer to two others presented by the same donor, and at the same time. In both cases, however, there is no incised sculpture. One is a linear-oval flat pebble, four inches long and two wide, shaped generally like a "Bull-roarer." It is similarly ruddled, and the smaller end is covered with gum cement on both sides. The second specimen is a flat shale pebble, broad-oval in shape, eight and a-quarter inches long, by four and a-half wide, unincised, but profusely ruddled on both aspects, although the ruddle is lighter in colour than in the incised implement (Pl. i., Figs. 1 and 2. The surfaces are speckled over with magenta coloured dots or spots.

With regard to the use of these implements I can do no more than offer a surmise.—The care bestowed on the spiral and circular figures on the one hand, and the thick coating of colour on the other, at once place on one side the supposition that they might have been used for grinding stones, for which purpose the shape and size of at least two, would admirably adapt them. Again, the entire absence of scratches favours this view. The general likeness to a "Bull-roarer" of the smallest and longest of the three pebbles, and the similarity of the sculpture of the incised implement to the circular ornament seen on some of these objects of Aboriginal veneration leads me to infer that these stones were employed in some of the Black's secret rites, but the precise use must still remain unknown.*

One of the most beautiful examples of circular concentric sculpture with which I am acquainted is represented in Pl. i., Figs. 3 and 4, a "Bull-roarer" from the Urania Tribe, Linda Creek, W. Queensland. There are five circular figures on each aspect of the implement, merely differing in size and the number of contained circles, and similar to those seen on the stone implement already described; they occupy more or less the entire surface of this implement.

The central and largest disk (Pl. i., Fig. 4), in which there are sixteen circles, is separated from the others by a crossbar above and below it, each of four incised lines; these do not occur on the other or slightly convex face of the "Bull-roarer." The uppermost and smallest disc on this aspect (Pl. i., Fig. 3) differs from the others in that the concentric circles are fewer in number, leaving a plain and unincised intermediate area between the outer circles and a central nucleus of three. Between this disc and the second, and below the fifth, are two incised arcs of four and three lines respectively, and similar to that already described on the stone implement (Pl. i., Fig. 2). This "Bull-roarer" is sixteen inches long by two and a quarter wide, and is more acutely pointed at one end than the other. It is attached to a long cord composed of human hair and fine emu down, and is covered with ruddle and grease.

The second and third "Bull-roarers" are equally well incised with circular and other figures. They are said to be from South Australia, but are, I think, more likely to come from Central Australia. Taking the larger one first, measuring fourteen inches by two inches, we see on the more convex of the two faces (Pl. ii., Fig. 5), a central figure answering to the uppermost in Pl. ii., Fig. 3, a nucleus of circles within a circumferential set, five in both cases. Above and below this is an arc or semicircle

---

* Since the above was written I have read the following passage in the "Horn Scientific Expedition Report" (Vol. i., Narrative, &c., 1896, p. 35), by Prof. Baldwin Spencer. Speaking of the *Churina* or "Bull-roarers," met with in Central Australia, he says—"Stone ones are still more valuable and sacred than wooden ones, which are usually spoken of as "Irula," the patterns on which are copied from the older stones, the history and origin of which are lost in the dim past." This rather tends to confirm the view I have taken of this incised stone implement.

similar to those already described in the first "Bull-roarer," and
the stone implement.

On the flat side of this implement (Pl. ii., Fig. 6), the carving
is very remarkable, consisting of indiscriminately scattered small
circles, and arcs or semicircles in various degrees of completeness
and position. Here and there are transverse short incised lines
proceeding from the margins inwards, precisely as the larger
incisions drawn in Pl. i., Figs. 1 and 2. These crossbars or transverse incisions are also seen in one of Hardman's figures* of the
Kimberley implements. On the convex face of the smallest
"Bull-roarer" (Pl. ii., Fig. 8) are four discs, each one surrounded
by two semicircles of concentric incisions, whilst the third from
the top is separated off by crossbars. On the reverse of this
implement (Pl. ii., Fig. 7) the ornamentation is again different,
consisting of a central longitudinal serpentine figure looped on
itself at the upper end, margined by bow-shaped figures of three
or more incisions, and the re-entering angles between the latter
occupied by short transverse bars. The execution of the incised
sculpture on this beautiful little implement is of a much more
finished nature than that on the preceding "Bull-roarer" (Pl. ii.,
Figs. 5 and 6), and more akin to that of the first described
(Pl. i., Figs. 3 and 4). It is ten inches long by one and a quarter
inches wide.

The question of this circular ornamentation or pictography
seems to have engaged the attention of writers on the Australian
Aborigines but little. It has been suggested by Mr. D. Brown,
who obtained examples from Stuart's Creek, Central Australia,
that these concentric rings indicate the practice of sun worship
on the part of those who carved them.† On the other hand,
Prof. R. Tate rejects the view that they are symbols at all, and
believes the execution of them to be merely a matter of sport.‡
He further very much doubted if they could be regarded as the
production of the untutored Aboriginal. It is, however, a curious
coincidence that one of the principal localities for these circular
inscised "Bull-roarers" is Kimberley, where at the time of Mr.
Hardman's explorations the Blacks had come in contact with the
White-man possibly as little as anywhere. Without entering
into the question of sun worship, although some of our Aboriginal
tribes seem to have possessed customs and practices suspiciously
like this form of adoration, even if they were unacquainted with,
or had lost their esoteric meaning—it may be pointed out that
the only published objects bearing this circular ornamentation
are "Bull-roarers," and as everyone knows these are the most

* Proc. R. Irish Acad. (2), i., 1888, No. 1, t. 3, f. 2.
† Trans. Roy. Soc. S.A., iii., 1880, p. xxiii.
‡ Trans. Roy. Soc. S.A., iii., 1880, p. xxiv.

precious and sacred of the Black's possessions, and only used in the mysteries of the Bora. I think, therefore, that unless Prof. Tate's view can be supported by stronger evidence than mere opinion, it must be dismissed, whatever the real significance of this circular incised ornament may be.

Mr. W. W. Froggatt, when in Kimberley some years ago, paid considerable attention to the practices of the Aboriginal inhabitants. He observes* that during initiation "men are stationed round whirling flat oval sticks, on which are carved *curious symbols.*" The italics are mine.

We know that amongst some ancient peoples, and even amongst the remnant of some existing, the circle or disc was symbolic of the sun. Our acquaintance, however, with the beliefs and esoteric mysteries of the Aborigines is too limited to hazard a suggestion that the figures on the "Bull-roarers" and stone implements bear a similar reference—but it is possible.

One of Mr. Hardman's "Bull-roarers" bears five sets of concentric circles, separated by groups of vertical incisions, and horizontal marginal ones, as in our Pl. ii., Figs. 6 and 7. A second implement bears irregular concentric semicircles at the apices, one on each side, and four sets of quadrangular figures concentric within one another. Two of the implements now figured are said to be from South Australia, but the correspondence in every way with Hardman's Kimberley figures† causes me to suspect that they must in reality come from the same district, or at any rate high up in Central Australia.

A few other cases of circular ornament in Australia may be mentioned, such as the circles, and ovals as well, carved on the trees surrounding the larger circle of a *Bora* ground near Gloucester, N.S. Wales,‡ and the numerous figures found by Mr. Richard Helms, during the progress of the Elder Exploring Expedition from South to West Australia. On a cave-shelter pictograph at Areoeillinna Wells, S.A. § are several of these concentric circles in red. Mr. Helms says these "are of very frequent occurrence, and have undoubtedly a symbolic meaning." Others were met with at Wa-Wee Rock Holes in another Cave-shelter, and at Mount Illibillie on white pigment. ‖ The most complete ones, however, were found in a similar situation near "Camp 6," Everard Ranges. Here is a circle in red of seven rings, a black nucleus, and radial bars passing from the centre

* Proc. Linn. Soc. N.S.W., (2), iii., 1888, Pl. 2, p. 652.
† Proc. R. Irish Acad. (2), i., 1888, No. 1, t. 2, f. 4-5A, & t. 3.
‡ Fraser; Aborigines of N.S. Wales, 1892, pl. opp. p. 11.
§ Trans. Roy. Soc. S.A., xvi., 1896, Pl. 3, t. 9.
‖ *Loc. cit.*, t. 10A & 11.

to the circumference.* The circular incised sculpture is very common on many petroglyphs, particularly in America, such as Bald Friar Rock, in Maryland; Girao, in Brazil; Cipreses, in Chili, and on the Colorado River, Utah,† and it is certainly curious to find this form of ornamentation whether on implements, as pictographs on the walls of Cave-shelters, or as petroglyphs, so widely distributed. It is curious and even startling to find the close general resemblance there is between this circular and spiral incised ornament on our Black's weapons, and in their Cave-shelters, and those curious petroglyphs found in odd quarters of the globe, and known as "cup-sculptures," both with and without a radial groove. Many of these were described by the late Mr. George Tate, occurring on Northumbrian (England) rocks, both circles and ovals, mostly with a radial groove.‡ Mr. Tate regarded them as the work of a Celtic race, and "symbolical most probably of a religious nature." Dr. B. Seemann has figured precisely similar closed concentric circles from the rock surfaces in Veraguas, New Granada, and believes them to have been produced by a very ancient people of that country, and to be "symbols full of meaning" to those who executed them.

I have lately seen a number of single circles on the petroglyphs of the Hawkesbury country around Narabine Lagoon, between Manly and Pittwater, both separately incised and forming portions of compound figures.

A SPEAR with INCISED ORNAMENT from ANGELDOOL, NEW SOUTH WALES.

By R. Etheridge, Junr., Curator.

A remarkably ornamented spear has been received from Angeldool, on the Narran River, by Dr. James C. Cox, who has been kind enough to present it to the collection. It is made from a sapling of light coloured hardwood, eleven feet nine inches long and two and a-half inches in its greatest circumference, tapering at both ends to a point. Unlike a very large number

---

* *Loc. cit.*, t. 13.
† Mallary; 10th Rep., Bureau Ethnol., U.S., 1893, pp. 86, 120, 153, 160.
‡ Tate; Anthrop. Review, iii., p. 293.

of Aboriginal spears, it is in one piece, and not with the head separately formed, and lashed or cemented on. I take it to be a hand-thrown weapon, and not propelled with the assistance of a womerah. The head of the spear, for eight and a-half inches from the apex, is blackened, then five alternating white and black bands follow occupying in the aggregate one foot, three of the bands white and two black. From this point downwards, to within nine inches of the proximal end, are six serpentine, but not encircling, continuous grooves, each bearing a series of close, backwardly directed, incised barbs, or teeth, and rendered prominent by having been coloured black. Spears similarly banded at the apex have been figured before, but neither Angas, Eyre, Wood, Smyth, or Knight, in their respective works, have given an illustration of one similarly ornamented with incised sculpture or decoration. With the exception of this feature, it is one of the type of such simple spears as the *Uwiuda*, of the Murray River,* or the *Koy-yun*.† Mr. E. M. Curr, however, states‡ that the Blacks of Hinchinbrook Island, and the adjacent mainland used carved spears, but he does not give particulars.

Smyth figures a simple spear with the distal end, or apex, segmented by white and black bands from West Australia,§ but otherwise it completely differs from the present weapon.

AN ACTINOCERAS from NORTH-WEST AUSTRALIA.

By R. ETHERIDGE, Junr., Curator.

(Plate iii.)

I am not aware that this interesting genus has so far been recorded from the Carboniferous rocks of West Australia. A rather fine example exists in our collection from the Lennard

---

* Angas ; S. Australia Illustrated, 1816, t. 51, f. 34.
† Smyth ; Aborigines of Victoria, i., 1878, p. 307, f. 83.
‡ Australian Race, ii., 1886, p. 418.
§ Smyth; *loc. cit.*, p. 337, f. 143.

River, less the oldest and youngest chambers of the shell, and unfortunately it has been crushed, more particularly in the upper portion of the specimen. The length is six inches, and there are within this space nineteen or twenty chambers, the upper with a breadth of one-quarter of an inch, and the lower a trifle less. At both ends the large beaded siphuncle is visible, above in the round, below in partial cross-section.

The siphuncle is nearly marginal in position, or in a perfect specimen would probably be sub-marginal. At the younger end it stands out from the crushed and partially denuded shell exhibiting portions of three of the "beads," or segments, so characteristic of the genus. The diameter in its present condition is nine-sixteenths of an inch, but at the older or lower end of the shell it is only three-sixteenths. The siphuncular segments to the naked eye are grooved and ridged, and where not abraided, the ridges are very slightly convex. An examination of the partially and naturally sectioned siphuncle at the older end of the shell, as well as in a cut section, reveals the fact that these grooves are the infolding of the siphonal membrane, as described by Mr. A. H. Foord* who says: "The calcified lining membrane of the siphuncle is thrown into a series of folds, which impart to it a puckered appearance, which is very characteristic." The same Author also observes that the shelly covering of the siphuncular segments, or "beads" composed of several layers, is very rarely preserved, but at the oldest end of the present specimen it is distinctly visible. Some good figures of the infolding of the membrane are extant, and foremost amongst these may be mentioned *Actinoceras Bigsbyi*, Stokes, as represented by Barrande.† In some of the infoldings, the membrane seems to expand into vertical sac-like cavities protruding inwards. When subjected to microscopic examination, in a thin section, the inflected portions of the siphonal membrane are seen to be comparatively thick, each one increasing slightly in width as it proceeds inwards, becoming somewhat truncheon-shaped, leaving in the centre a narrow free space filled with impalpable matrix. They are variable in length, some long, some short, but never approaching the centre of the siphuncle. At the point through which the section is taken there are seventeen of these inward prolongations, but they do not appear to be developed with equal regularity as to distance apart around the rather oval siphuncle. Furthermore, these prolongations appear to be open to variation in shape, for along one side are two assuming a decidedly pyriform outline, and a third that seems to show signs of bifurcation at its inward end, although too much stress must not be laid on this point. There is no trace of the endosiphon,

---

* Cat. Foss. Ceph. Brit. Mus., 1888, Pt. i., p. 166.
† Syst. Sil. Bohême, ii., t. 231.

nor remains of its tubuli. The chambers are narrow, about four-eighths of an inch in the upper portion and three-eighths of an inch in the lower portion of the shell. There are four and six septa to the inch respectively in the parts referred to, increasing very slowly in their distance apart, and with plain edges. The siphuncle is a good deal inflated between the septa, wider than long. The external shelly-layer is not preserved, and in consequence the sculpture is not known.

I propose to call this species *Actinoceras Hardmani*, in honour of the late Mr. E. T. Hardman, who acted as Geologist to Forrest's Kimberley (N.W. Australia) Exploring Expedition in the years 1883-84, but who was perhaps better known through his connection with the Geological Survey of Ireland.

THE DISCOVERY OF BONES AT CUNNINGHAM CREEK, NEAR HARDEN, N.S. WALES.

By R. ETHERIDGE, JUNR., Curator.

The Cunningham Creek Gold-field is situated about fourteen miles south-east of Murrumburrah and Harden. The "diggings" lies along both sides of the creek, above and below the Jugiong Road—crossing to Cunningham Plains, reaching almost down to its junction with the more important Jugiong Creek. The whole of this district is composed of grey granite cropping out here and there in bosses and tors, otherwise a thick granitic detritus hides the bedrock completely, and in consequence a subsequent denudation has given rise to gently rolling downs and hills. It is in this detritus that the bones of extinct Marsupials have been found for some time past, generally lying immediately above the auriferous wash-dirt of the old subsidary branches of Cunningham Creek. The claim of Messrs. J. F. Wilson and Party, who first reported the discovery, is situated on the north bank of the creek, the shaft mouth being about seventy feet above the creek bed, and on the Cunningham Creek Common, barely a mile south-west of Cahill's Hotel. The shaft is down sixty feet in fine granitic detritus, interspersed with large boulders of granite. The bones are usually met with at fifty-eight feet from the surface, and, as before stated, immediately above the wash-dirt, but from the wet nature of the ground, they are all very rotten, and difficult of extraction and preservation. The wash-dirt appears to be of poor quality, although containing a few gem-stones, running in narrow gutters between hard granite bars. The bones procured were chiefly those of *Diprotodon*.

Through the courtesy of Mr. W. T. Ditchworth, the Manager of the Crown Point Gold Mining Co., Ltd., I was able to inspect the workings of the Marshall-McMahon Reef, where a quartz lode carrying free gold, and another with very refractory ore, are worked. I was fortunate enough to obtain good specimens for our collection.

ADDITIONAL LOCALITIES FOR *PERIPATUS LEUCHARTII—Säng.*

BY THE LATE FREDERICK A. A. SKUSE, Entomologist.

The writings resulting from the researches of Dendy, Spencer, Fletcher and others, have for some time past aroused considerable interest in *Peripatus* in Australia ; so that every scrap of additional information respecting these remarkable creatures may be considered of some value, and the evident interests attached to a new discovery affecting our knowledge of *Peripatus* lends no mean impetus to its investigation and the seeking out of its distribution.

During a recent visit (Oct. 22nd, 1895) to Colo Vale, near Mittagong, N.S.W., Mr. Edgar R. Waite* chanced upon a specimen of *P. leuchartii*† whilst searching beneath fallen timber for reptiles and insects. Colo Vale lies on the Great Southern Railway line, seventy-two miles from Sydney, and the specimen of *Peripatus* was obtained at an altitude of 2,000 feet.

Other examples have just been presented to the Museum by Mr. C. J. McMasters, who obtained them at Moree, New South Wales, and plentifully by the Curator in November, 1895, in and under rotten logs in the vicinity of the Jenolan Caves, Blue Mountains, New South Wales, at an altitude of 4,000 feet.

---

* Waite ; Proc. Linn. Soc. N.S.W. (2), x., 1895, p. 519.

† Fletcher ; Proc. Linn. Soc. N.S.W. (2), x., 1895, p. 183, considers "all the known Australian specimens of *Peripatus* as referable to one comprehensive species, *i.e.*, *P. leuchartii*., Säng.

It might here be mentioned that in 1887 **Prof. Jeffrey Bell** contributed a note* on the "Habitat of *Peripatus leuchartii*," wherein he mentions that previously the place of origin of this species was vaguely stated as "New Holland" and on the receipt of two specimens from Dr. E. P. Ramsay, of Sydney, gave the Queensland scrubs, near Wide Bay, as the more precise locality. The error is perpetuated by Sedgwick,† who incidentally remarks, "the finder's name has not been communicated to me."

I would point out that the specimens referred to as coming from Wide Bay were collected by myself on April 3rd, 1887, when I obtained several examples under stones close to the Hospital and Acclimatisation Society's Grounds, in Brisbane. These were handed to Dr. Ramsay, who sent two specimens of them to Prof. Bell for study. They were ultimately forwarded to Prof. Sedgwick for inclusion in his Monograph.

Mr. Henry Tryon previously recorded‡ the finding of other examples from the same locality in conjunction with myself, but their identity with those mentioned by Sedgwick has not to my knowledge previously been made known.

Mr. Chas. Hedley tells me that he found *Peripatus* under a log by the road-side at the altitude of 2,000 feet in 1889 at Cunningham's Gap, South Queensland, and that this specimen was pronounced by Mr. Tryon, then of the Queensland Museum, to be *P. leuchartii*, which determination was no doubt correct.

DESCRIPTION of a NEW PAPUAN LAND SHELL.

By C. Hedley, Conchologist.

Thersites septentrionalis, *n. sp.*

Shell turbinate conic, narrowly perforate, large, solid, brownish-yellow with deep chocolate bands. A third of the base is occupied by a broad chocolate band whose outer margin reaches the insertion of the lip, a yellow peripheral zone of less width follows, a chocolate band as wide as the last and which becomes supersutural in the upper whorls, a narrow yellow, a wider chocolate, a narrow yellow, and a narrow subsutural chocolate

---

*Ann. & Mag. Nat. Hist. (5), xx., 1887, p. 252.

†Qt. Journ. Micro. Sci., xxviii., 1888, p. 431.

‡Proc. Roy. Soc. Qd., iv., 1887, p. 78.

band then successively occur. The sculpture consists of fine raised growth lines which slightly pucker the suture; the reticulations characteristic of many Queensland species were not visible in the specimens under examination, from which they might, however, have been worn. Suture impressed. Whorls five, convex, descending rather suddenly at the aperture for the breadth of the yellow peripheral zone. Aperture oblique, squarish, light within and showing the chocolate bands. The lip is blackish, very glossy, thickened, and widely reflected throughout, the columellar expansion almost covering the deep narrow umbilicus. A thin transparent callus extends from insertion to insertion of the lip.

Length 52 mm. Breadth 38 mm.

Collected on the Musa River, on the North East Coast of British New Guinea, by His Honour Sir W. MacGregor, K.C.M.G.

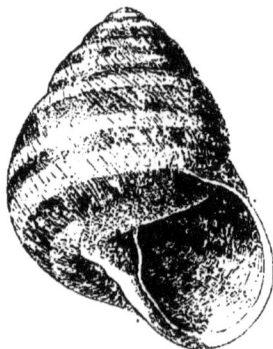

The original of this description and life size figure is registered in the collection of the Australian Museum as C. 2890.

This species, and *T. broadbenti*, Brazier, I now consider to be the only representatives of *Thersites* yet discovered in New Guinea. Much interest therefore attaches to so handsome and characteristic a species as the novelty occuring in a locality so remote from the principal seat of the genus. Though differing in size and colour, an unfigured Queensland species *T. etheridgei*, Brazier, closely approaches in contour and perforation, and may be held as nearest in systematic order.

Specimens were received, perforated, and strung together in bunches for native ornaments. One tassle contained this species and *Chloritis rehsei* tied together. The animal has not yet been seen.

ON A NEW SUB-SPECIES OF *PSOPHODES CREPITANS*.

BY ALFRED J. NORTH, C.M.Z.S., Ornithologist.

For many years past I have known that the specimens of *Psophodes* collected by Messrs. Cairn & Grant in 1887 and 1889 at Boar Pocket, North-eastern Queensland, were different in several respects from the *P. crepitans*, inhabiting New South Wales and Victoria. While lately examining the Reference Collection I found another skin from the same locality, which was obtained by Mr. W. S. Day on the 4th of May, 1891. As all the specimens collected at Boar Pocket at wide intervals are alike, I take the present opportunity of pointing out the distinctive characters of this northern race of *Psophodes crepitans*.

PSOPHODES CREPITANS LATERALIS, *subsp. nov.*

Adult.—Like *P. crepitans*, Gould, but differs in having the lateral feathers of the tail, which is shorter, tipped with pale-brown instead of white. Moreover, the sides of the lower flanks are olive-green, whilst in *P. crepitans* they are ashy-brown. Total length 9·8 inch; wing 3·9, tail 5·4, bill 0·9, tarsus 1·3.

Hab.—Boar Pocket, N.E. Queensland.

Type.—In the Australian Museum, Reg. No. O·4645.

Remarks.—I have selected the largest and finest specimen as the type. The tail measurements of two more adult males from the same locality are respectively 5·2 and 5 inches. Specimens of *P. crepitans*, obtained in the Illawarra District of New South Wales vary in the total length from 10·5 to 11 inch. and in the length of the tail from 5·75 to 6·1 inch.

Boar Pocket is situated on the table-lands of the Upper Barron River, about thirty-two miles from Cairns.

ORNITHOLOGICAL NOTES.

By Alfred J. North, C.M.Z.S., Ornithologist.

——◆——

I.—On the EXTENSION of the RANGE of *CALAMANTHUS FULIGINOSUS*, and *EMBLEMA PICTA* to NEW SOUTH WALES.

During the months of August and September, 1896, the Ornithological Collection of the Australian Museum has become enriched by the receipt of specimens in the flesh of *Calamanthus fuliginosus*, and *Emblema picta*. The former species was obtained on Boloco Station, near Buckley's Crossing Place, New South Wales, on the 19th August by Mr. E. Payten, who killed it with a stone. It was then taken to Mr. Reuben Rose, the owner of the station, and was by that gentleman presented to the Trustees of the Museum. Buckley's Crossing Place, situated on the Snowy River, is about 296 miles south of Sydney and 34 miles as the crow flies to the nearest point of the imaginary line between Cape Howe and Forest Hill, which separates the south-eastern corner of New South Wales from Eastern Victoria. The natural or artificial boundaries of the Continent of course do not form any barriers to birds, but hitherto *C. fuliginosus* has been recorded only from the southern parts of Victoria and South Australia; Tasmania being the stronghold of the species. From typical examples of *C. fuliginosus*, the bird procured in New South Wales differs in the following respects: the bill is shorter, the throat is buff instead of white, although similarly streaked with black, and the outer webs of the primaries are externally edged with ashly-white. These slight differences may be due to immaturity, or climatic variation, the locality in which it was obtained being over 2,000 feet above the level of the sea.

On the 23rd of September, Mr. A. M. N. Rose presented to the Trustees three adult male specimens of *Emblema picta* in splendid plumage. These birds were shot the previous day by his nephew, Mr. Arthur Payten, at Campbelltown, an agricultural and dairy-farming district, 34 miles south-west of Sydney. Mr. Payten saw altogether five specimens, which kept together in a small flock while searching for grass-seeds on a hill devoid of any cover. Previously this rare bird has been recorded only from North-west Australia where the type was procured; from Derby and Cambridge Gulf by Mr. E. J. Cairn and the late T. H. Boyer-Bower, and from several localities in Central Australia,

where it was obtained by the members of "The Horn Scientific
Expedition." Why this small flock should have wandered so far
south-east instead of pursuing the usual course of migration to the
north-west it is difficult to conjecture, unless the birds followed in
the track of an abundant rainfall so common to Central Australia,
with its rapidly accompanying growth and profusion of rich
grasses, thereby causing a plentiful food supply. There is no doubt
whatever that reaching the western border of New South Wales
the excessively dry season now being experienced there has
driven this small nomadic flock from the withering and burnt-up
grass lands to the cooler districts near the coast. This is only a
repetition of the effects of last year's drought when many birds
whose habitat is the dry inland districts of the Colony, were
obtained near Sydney, among which may be mentioned *Falco
hypoleucos*, a typical Central Australian species.

Roughly estimated, the nearest recorded locality in Central
Australia in which *Emblema picta* has been obtained, is 1,300
miles in a direct line from Campbelltown, in New South Wales,
where the present specimens were procured.

II.—ON A CURIOUS NESTING-SITE OF *ANTHUS AUSTRALIS*.

(Plate iv.)

The Trustees of the Australian Museum are indebted to Mr.
A. M. N. Rose, for a nest of the Australian Pipit or common
"Ground Lark," *Anthus australis*, placed in a very curious
position. It is built inside an old rusty preserve tin, measuring
four inches and a half in length by three inches and a half in
diameter. The entrance to the nest is narrowed to two inches,
by a small platform of dried grasses which protrudes out of the
mouth of the tin. This nest was found on the 24th of November,
1896, by Mr. A. Payten at Campbelltown in the same paddock
as he shot the specimens of *Emblema picta*, and contained two
slightly incubated eggs. The tin, which has the lid still attached,
but bent at a right angle, was lying exposed on the ground,
without shelter or concealment of any kind, beyond a few short
blades of dried grass. The eggs are of the usual type, a greyish-
white ground colour thickly freckled all over with pale brown
markings; length (A) 0·8 x 0·67 inch; (B) 0·84 x 0·67 inch.
As will be seen on reference to the accompanying plate, it is a
curious site for a bird to select which builds an open cup-shaped
nest concealed only by an overhanging tuft of grass, or the
surrounding herbage.

III.—On the NIDIFICATION of *MEGALOPREPIA MAGNIFICA*, the MAGNIFICENT FRUIT PIGEON.

Columbia magnifica, Temm., Trans. Linn. Soc. xiii. p. 125 (1821).
Carpophaga magnifica, Gould, Bds. Austr. v. pl. 58 (1848).
Megaloprepia magnifica, Salvad., Cat. Bds. Brit. Mus. xxi.
        p. 167 (1893).

The Magnificent Fruit Pigeon is freely dispersed throughout the rich coastal brushes of Eastern Australia, from the neighbourhood of Cairns in North-eastern Queensland to Berry in New South Wales. In the latter colony it is far more frequently met with in that rich belt of luxurious vegetation lying between the Tweed and Bellinger Rivers, than it is in the humid valleys and mountain ranges of the south coastal district. At Cairns it overlaps the closely allied, but decidedly smaller species *M. assimilis*, which ranges northward from that locality to Cape York.

Although *M. magnifica* is plentifully distributed throughout these brushes nothing has hitherto been recorded of its nidification. For an opportunity of examining a nest and egg of this species I am indebted to Mr. George Savidge, a most enthusiastic oologist, who has lately found this fine Pigeon breeding on the Upper Clarence. Mr. Savidge has also forwarded me a skin of the female shot at the nest, together with the following notes relative to procuring the nest and egg :—

" Having been told by some timber-drawers that they had discovered three nests of *Megaloprepia magnifica* at Pine Scrub, Oaky Creek, Upper Clarence, each with a single egg, and upon which the birds were sitting, I determined to pay a visit to these scrubs to search for the nests. Accompanied by a friend, Mr. Thos. Woods, and an aboriginal called Freddy, we started at daylight on the morning of November 8th, 1896, and arrived at our destination, twenty-five miles distant a little after ten. Several nests were seen but they contained neither eggs or young, and after a long search we decided upon going further into the scrub. The peculiar call of *M. magnifica* could be heard on all sides, and at last after searching for several hours we saw one fly from a tree about twenty-five yards in advance. Upon nearing the tree we discovered the nest, and the egg could be plainly seen in it. Wishing to obtain the bird we sat down for some time, but eventually decided to move lower down the creek into closer concealment, as I thought the bird might be watching us and would not return. After waiting a quarter of an hour we observed the Pigeon fly back and settle on a thick branch. I did not fire as it was a bit too far, and its body protected by the limb it was sitting upon. After having a good look round it flew into a thick patch of scrub a few yards away and was lost to view, however, it soon came back and settled about two feet from the nest, and facing us. I was afraid the spread of shot might shatter the nest, but as it was getting late I fired and the bird fell into a small pool of water beneath. The egg was secured after some trouble as the nest was built on the end of a thin outspreading branch of a 'Scrub Elm,' about twenty feet from the ground. The scoop had to be used, and the nest was so small I was afraid the egg would roll over, and it took the black some time before he got it safely into the net. The limb was then chopped off and the nest secured. Upon dissection of the bird, which proved to be the female, no other egg was found in it approaching maturity, the largest being the size of a pea."

The nest of *M. magnifica*, is an exceedingly small and perfectly flat structure, and with the exception of a few long straggling sticks lying almost parallel to the branch on which it is placed, barely averages five inches in diameter. It is built at the junction of a forked horizontal branch of an *Aphananthe phillipinensis*, which is partially covered with a growth of moss. The nest is composed of thin sticks and twigs intermingled with the wiry spiral tendrils of a vine; the latter material wholly forming the centre of the structure for the reception of the egg. When sitting the Magnificent Fruit Pigeon would almost conceal the nest for very little of it is visible below the branch. The green leafy twigs which sprout out in close proximity to the nest, also harmonises well with the colour of the back, wings, and tail of the sitting bird, and renders it less liable to detection. The egg is pure white, elongate oval in form, and there is very little difference in the shape of the two ends, the texture of the shell being very fine and the surface lustreless. It measures 1·57 inch in length by 1·2 inch in breadth.

Specimens of *Megaloprepia magnifica* and *M. assimilis* from different localities measure as follows :—

M. magnifica.

Sex.	Length.	Wing.	Tail.	Bill.	Tarsus.	Locality.
♀ ad.	17·25 in.	9·1	7·6	0·83	1·1	Clarence River, N.S.W.
♂ ad.	17·5	9·1	7·75	0·85	1·1	Maryborough, Wide Bay, Queensland.

M. assimilis.

Sex.	Length.	Wing.	Tail.	Bill.	Tarsus.	Locality.
♂ ad.	13·5	7·6	6·75	0·78	1·05	Cairns, N.E. Queensland
♀ ad.	13	7·1	6·3	0·62	1 in.	Cape York, N. Australia.

7th January, 1897.

An AUSTRALIAN SAUROPTERYGIAN (CIMOLIOSAURUS), CONVERTED INTO PRECIOUS OPAL.

By R. ETHERIDGE, JUNR., Curator.

(Plates v., vi., vii.)

I WAS recently favoured by Messrs. Tweedie and Wollaston, Merchants, of Adelaide, through the good offices of Mr. H. Y. L. Brown, Government Geologist for South Australia, with a large quantity of opalised material from White Cliffs, representing the broken-up skeleton of a Plesiosaur, but unfortunately wanting the skull. There are numerous vertebræ in various states of completeness, innumerable portions of ribs, a few teeth, phalanges, and other bones that will be subsequently referred to. These have now become the property of the Trustees of the Australian Museum.

I.—PRECIOUS OPAL AS AN AGENT OF REPLACEMENT.

The replacement of the calcareous matter in fossils by Precious Opal appears to be a fact but little commented on by Authors.

The search for opal in the Upper Cretaceous at the White Cliffs Opal-field on Momba Holding, about sixty-five miles north-north-west of Wilcannia, in Co. Yungnulgra, has been signalised by the discovery of many beautiful examples of the entire conversion of the shelly envelopes of Pelecypoda and Gasteropoda, the internal shells of Belemnites, and Reptilian remains, into Precious Opal by a process of replacement.

Many of these are in the Collection of the Geological Survey of N.S. Wales, others have been lent to the same, and through the courtesy of Mr. E. F. Pittman, Government Geologist, I have been permitted to examine them.

The process of "silicification," as it is called, or the replacement of matter in fossil organic remains, by silica, in one or other of its varieties, is too well-known to require more than the briefest notice.

Silicification is said to be *primary* when organisms have undergone a slow process of alteration in water holding silica in solution, each particle of tissue, as it decayed, being replaced by the mineral in question, the minute structure of the body thus acted on being so preserved. "By far the commonest mode of replacement is that whereby an originally calcareous skeleton is replaced by silica. This process of 'silicification'—of the replacement of *lime* by *silica*—is not only an extremely common one,

A

but it is also a readily intelligible one ; since carbonate of lime is an easily, and flint a hardly soluble substance. It is thus easy to understand that originally calcareous fossils, such as the shells of Mollusca, or the skeletons of Corals, should have in many cases suffered this change, long after their burial in the rock, their carbonate of lime being dissolved away, particle by particle, and replaced by precipitated silica, as they were subjected to percolation by heated or alkaline waters holding silica in solution."* On the other hand, if the minute structure of the fossils has been injuriously affected during this process, or destroyed, notwithstanding the preservation of the outward form, the silicification is said to be *secondary*, having taken place at a period long posterior to the entombment of the organic remains.   "In the first stage of the process," adds Prof. H. A. Nicholson, from whom I am quoting, "the outer layer of the fossil very commonly becomes converted into, or covered by, small circular deposits of silica, having the form of a central boss surrounded by one or more concentric rings ('orbicular silica,' or ' Beekite markings '). If the process goes on the whole of the fossil may ultimately become converted into flint."

A *third* form of silicification may, I believe, exist—the conversion of the original calcareous matter into the form of chalcedony, so excellently seen in the shells (*Physa*, etc.) of the Lower Intertrapean chert beds of the Deccan Tertiary Trap Series at Nagpur, in India, or the chalcedonic Permo-Carboniferous Brachiopoda of Point Puer, Port Arthur, Tasmania.

The mode of occurrence of the Opal at White Cliffs has already been so fully described by Mr. W. Anderson and Mr. J. B. Jaquet that it need only be briefly referred to. It is met with in beds of kaolin and conglomerate forming a portion of the Desert Sandstone, but the former author also says in the "vitreous-looking" Desert Sandstone itself. Four separate conditions of occurrence are detailed† by Mr. Jaquet, viz. :—

1. In thin horizontal veins, between the bedding planes of the kaolin.

2. As irregular nodules scattered through the kaolin.

3. As Wood Opal.

4. As opaline shells, etc.

We are at present concerned only with the two last.

The Wood Opal is usually of an opaque milk-white or horn-yellow colour, and is simply hydrous silica, although the woody structure is still visible and in some instances well preserved, but in other

* Nicholson, Man. Palæontology, 3rd edition, i., 1889, p. 7.

† Ann. Rep. Dept. Mines and Agric. N.S.W., 1892 [1893], p. 141.

specimens hardly any structure at all is to be observed, beyond the outward form of stem or branch, as the case may be. Not infrequently radial cracks are filled with Precious Opal, when the play of colour is very fine.

The Animal Remains occur under the following conditions :- -

1. As external or internal casts in kaolin, without opalization of any kind.

2. Entirely converted into hydrous silica or Common Opal, white and opaque, but occasionally with traces of the coloured variety scattered through.

3. Wholly or partially converted into translucent-glassy to vitreous semi-opaque Precious Opal, displaying a fine range of colour.

The colours visible by reflected light are principally blue, red, green, and yellow, with their various shades and combinations, not the least pleasing being an ever-varying degree of red and blue-tinted purple.

When the fossils are in the form of kaolin casts, specific identification, with a very few exceptions, is almost unattainable. Those in which opalisation, however, has taken place, are always determinable, more or less, and the substitution of the original carbonate of lime has been very thoroughly carried out. Fragments of these opalised remains, chiefly shells, are freely scattered throughout some hand-specimens of the opaline kaolinised conglomerate, from the bed B of Mr. Jaquet's section.* The kaolin casts are either white or tinged with iron oxide, arising from the highly ferruginous clays that Mr. Jaquet says the kaolin passes into.

The opalised fossils comprise Crinoid remains, the shells of Pelecypoda and Gasteropoda, portions of Belemnite guards, and Sauropterygian bones. The preservation of some of these fossils is excellent, although all are not alike in this respect, and the extent to which the opalisation has at times been carried is remarkable. In some Pelecypoda, the external growth laminæ, and intermediate sculpture striæ are fully preserved, whilst the shell substance is completely changed, and by transmitted light the valves of many are almost transparent. On the fractured edges of one of these bivalves the glassy opal is quite translucent by reflected light. When such valves are met with in apposition, the interiors are often found to be filled with soft kaolin, and no better examples of the complete change that has taken place can be examined.

The replacement of the fibrous calcite of the Belemnite guard, when viewed in cross-section, presents a far less translucent, and

*Ann. Rep. Dept. Mines and Agric. N.S.W., 1892 [1893], p. 141.

much more opaque and vitreous-looking appearance than that seen in the other Mollusca. In one small guard in particular, now before me, remains of the radiating fibres and concentric layers of calcite are visible round the periphery, gradually fading off into a dark blue and purple vitreous-looking opal.

Pre-eminent for its beauty is a bivalve, obligingly lent to the Geological Survey of N. S. Wales for examination by Mr. H. Newman, jeweller, of Melbourne. This is without exception one of the most beautiful conditions of fossilisation I ever beheld— perfectly clear of the matrix, with the valves in apposition, and save for a slight crushing about the centre of one of them, quite perfect, wholly converted into Precious Opal, and with a play of colour quite equal to the fragments in quartzite shortly to be referred to. The shell substance is almost glassy transparent. It is probably identical with the shell already referred to, with the translucent fractured edges, from the collection of Mr. G. de V. Gipps, also lent to the Geological Survey.

Mr. J. E. Carne informs me that the Survey Collection contained, previous to the Garden Palace fire, an Ammonite, wholly converted into Precious Opal, six inches in diameter! This came from White Cliffs, and was probably one of the first fossils ever obtained there.

By no means the least interesting specimen found in this field, previous to Messrs. Tweedie and Wollaston's reptile, is the half, split longitudinally, of a Sauropterygian vertebra, with the osseous matter converted in the first instance into the common white and opaque opal, and the canals and lacunæ remaining open and filled with a little ferruginous powder. The roughened edges of the fractured surfaces are then tipped, and the cavities to some extent lined with Precious Opal. This is also from the cabinet of Mr. J. de V. Gipps.

Some polished hand specimens of a highly fossiliferous chocolate-brown quartzite were presented by an unknown donor to the Geological Survey Collection, the whole of the organic remains being converted into Precious Opal, and the interstices between the component constituents of the rock likewise similarly filled as a secondary infiltration, probably replacing the calcite particles of the deposit. Beyond the fact that these specimens come from White Cliffs, I am not in possession of information as to the stratigraphical position of this quartzite, but possibly it may be derived from the water-worn vitreous boulders mentioned by Mr. Jaquet as occurring in the clay and conglomerate beds. The organic remains are those of Mollusca, with traces of Corals and stem-joints of Crinoids, showing such a marvellous kaleidoscopic play of colours that words are quite lacking to render the general appearance of the specimens appreciable.

There is a univalve in these hand specimens after the Euomphaloid type, and in consequence of the direction in which the latter have been cut, the sections are almost invariably across the whorls. It is a small shell, the largest not measuring more than three-eighths of an inch in diameter, biconcave, and the inner whorls barely distinguishable. From the difference in outline exhibited by the cross-section and the body whorls, I should say two, if not three species are present. The next commonest section is probably that of a Brachiopod, very geniculate in outline, the sections passing from the umbo to the front margin of either valve. One section in particular catches the eye from the comparatively large process, projecting from underneath the incurved umbo. This may be either a fulcrum supporting some of the internal shelly plates of a Brachiopod, or a spoon-shaped cartilage process of a Pelecypod, but I am inclined to the first opinion. Several small circular bodies, hollow in the centre, about three-eighths of an inch in diameter, and with median central vacuities, are scattered at random through the rock. There is no definite structure observable in these, but the size and general appearance closely resemble that of the stem joints of many Crinoids.

I am strongly of opinion that these highly opalised chocolate-coloured quartzites are of Devonian age, being portions of travelled blocks, in all probability coming from Mr. Jaquet's bed C.

The whole of the specimens now under discussion, were submitted to a careful examination by Dr. Thomas Cooksey, Mineralogist to the Australian Museum, and myself. Dr. Cooksey is of opinion[*] that in the chocolate quartzite the carbonate of lime of the fossils has been in the first instance converted into crystalline calcite, and the latter then replaced by secondary silicification in the form of Precious Opal. The traces of the cleavage planes and twinning of the calcite crystals are still preserved in the opal, the former in a great measure serving to produce that play of colour which gives to the opal its beauty and value. A few instances of a similar process are certainly visible in the opalised shells from the kaolin deposit, but in the majority of these there appears to have been simply a secondary replacement by hydrous silica of the ordinary carbonate of lime of the Molluscan and other tests.

There is no trace amongst these fossils of the Beekite stage of silicification. The occurence of this mineral in connection with "Fossil Organic Remains" in N.S.Wales is by no means an uncommon one, and has already been noted by the Writer.[†] Such occurrences, however, are confined to some of our Silurian and Permo-Carboniferous fossils.

---

* Rec. Aust. Mus., ii., 7, 1896, p. 111.
† Ibid, ii., 5, 1898, p. 74.

II.—THE REPTILE REMAINS.

To return to Messrs. Tweedie and Wollaston's reptile, *Cimoliosaurus leucoscopelus*, the remains that can be satisfactorily determined from the mass of opalised material are as follows :—

17 Cervical vertebræ, usually fragmentary.
2 Humeri.
4 Teeth.
Rib fragments (numerous).
Phalanges (numerous).

The absence of the skull practically bases systematic determination on the vertebræ and teeth, hence the following facts in the first place lead to the conclusion that it is a Sauropterygian :— (1) the vertebræ are amphicœlus ; (2) the single costal facets of the cervical vertebræ are entirely on the centra ; (3) each rib articulates to a single vertebra ; (4) the epiphyses of the humeri are enlarged ; (5) the teeth have more or less curved and sharp crowns with fluted enamel.

In the second place the form of the cervical vertebræ, and the absence of a foramen in the humeri indicate the Plesiosauridæ as the family to which these remains should be referred ; whilst within this family the complete anchylosis of the neural arches and cervical ribs to the centra, single costal facets on the cervical vertebræ, and the slender and non-carinated teeth point out *Cimoliosaurus*, Leidy, as the natural resting place of this reptile from White Cliffs Opal-field.

The cervical vertebræ are much elongated between their dorsal and ventral extremities. They vary in size, as usual, according to their position in the whole series, but the measurements of three of the more perfect and typical are as follows :—

	1.	2.	3.
Length of centrum body	1 in.	$1\frac{1}{16}$	1 in.
Height „ „	$1\frac{2}{8}$	$1\frac{7}{16}$	$1\frac{7}{16}$
Broadth „ „	$1\frac{1}{2}$	$1\frac{11}{16}$	$1\frac{4}{8}$
Height, including neural spine	$3\frac{1}{2}$	...	...
Breadth of cup	$\frac{7}{8}$	$1\frac{1}{16}$	$1\frac{5}{16}$
Fore and aft extent of neural spine at about the middle	$\frac{5}{8}$	...	...
Fore and aft extent of the neural arch below the zygopophyses	$1\frac{1}{8}$	$1\frac{3}{16}$	$1\frac{9}{16}$
Fore and aft extent of costal surface ...	$\frac{3}{8}$	$\frac{3}{8}$	$\frac{5}{8}$
Height of neural canal	$\frac{7}{16}$	$\frac{7}{16}$	$\frac{1}{2}$
Breadth „ „	$\frac{7}{16}$	$\frac{9}{16}$	$\frac{5}{8}$

The measurements are in inches and parts of inches.

The centra in these vertebræ are decidedly short, more proportionately so than in *Cimoliosaurus cantabridgiensis*, Lydk.,[*] and in this respect approach nearer to those of *C. valdensis*, Lydk.,[†] and *C. eurymerus*, Phill.[‡] There is not the slightest appearance of any rugosity round the edges of the terminal faces of the centra as in *Plesiosaurus rugosus*, Owen,[§] from the Lias, but they are prominent and outwardly bevelled as in *C. cantabridgiensis*; nor is there any sinuous profile with overhang of the upper border, and prominence of the lower border of the centra as in the genus *Polyptychodon*. The sides of the centra can hardly be described as concave, although the ventral surfaces are fairly so on either side the hæmal carina. The anterior and posterior articular surfaces vary in contour from circular to subquadrate, the transverse diameter being always the greater, with a well marked although not thick border, surrounding a wide and fairly deep concavity or cup. There are no mammillæ, or any trace of a pit in the vertebræ examined. The venous or hæmal foramina are situated in definite depressions, well marked and deep, and in the best preserved vertebra No. 1, (Pl. v., Fig. 4) three-sixteenths of an inch apart, this being the transverse measurement of the hæmal carina. The latter expands fore and aft in buttress formation into the anterior and posterior peripheries of the terminal faces, producing on the whole an hour-glass shaped figure, as in *C. constrictus*, Owen. The single costal facets are rarely seen in consequence of the thorough union that has taken place between the head of the ribs and the costal surface itself. In one, however (No. 2), where the head of the rib appears to have broken out, the pit or scar seems to be circular. The fore and aft borders of the neurapophyses are vertically concave, the fore much more so than the hind, and transversely are more angular than convex, particularly posteriorly. The neuro-central suture is almost totally obliterated. Zygopophysial ridges, the prominent lateral oblique ridges extending from the pre-zygopophyses to the posterior borders of the pedicle, can hardly be said to exist. The pre-zygopophyses are rather high, and in the only vertebra in which they are sufficiently preserved No. 1, (Pl. v., Fig. 5), do not project forward beyond the vertical line of the terminal face of the centrum, not even as much as in *C. valdensis*, Lydk., and *C. limnophilus*, Koken.[||] The zygopophysial articular surfaces approach the oval in form, and are very obliquely inclined, much more so than in either the last-named species, *C. cantabridgiensis*, or *C. eurymerus*. The post-zygopophyses are not preserved in the most perfect vertebra of the

---

* Cat. Foss. Reptilia and Amphibia Brit. Mus., pt. ii., 1889, p. 183, f. 60.

† *Ibid*, p. 188, f. 61.

‡ Lydekker, *ibid*, p. 206, f. 67.

§ Mon. Foss. Reptilia Liassic Formations, pt. i., 1865, p. 35.

|| Leydkker, *loc. cit.*, p. 225, f. 69.

collection, but broken off short at the point of their enlargement from the neural spine. The latter extends one and three-quarters of an inch from the roof of the neural canal upwards, and is five-eights of an inch in fore and aft measurement immediately above the zygopophyses, linear oval in section, and very sharp along the anterior and posterior edges; anteriorly this sharp edge is continued downwards over the neurapophysial surface as a similar ridge to between the prezygopophyses, and is but shallowly excavated, much less than in *C. cantabridgiensis*, nor anything as much inclined as in *C. valdensis*, or the extreme condition of the spine in *C. eurymerus*, but seems to approach nearest to that of *C. limnophilus*. The neural canal is large and subquadrate, proportionately larger in fact than appears to be general in this genus, even more so than in *C. limnophilus*.

The ribs are so reduced by fracture that it is almost impossible to afford precise details about them, but the majority appear to have been irregularly oval in cross-section, flattened on the sides. The costal articulations on the centra are invariably single.

A small portion of bone is figured (Pl. vii., Fig. 2) under the belief that it may be the expanded distal end of a short or cervical rib, forming one of the "hatchet bones." Notwithstanding its incompleteness, I believe this to be the nature of the specimen. The body is angular on the inner or hæmal surface, and flattened on the outer; the anterior extension is shorter than the posterior, itself abbreviated and thick. In Pl. vii., Fig. 4, is probably represented the distal end of one of the trunk ribs, broken at the apex.

Several interesting fragments are present, similar to Pl. vi., Fig. 5. I labour under the impression that these may be the expanded distal ends of the diapophyses of the dorsal vertebræ. They possess a distinct oblique convex articular facet, bounded by a slightly elevated ridge. Such portions of the shafts as remain are compressed between dorsal and ventral, and are angular transversely, *i.e.*, fore and aft. Pl. vii., Fig. 3, is another diapophysis, possibly one of the anterior with a portion of the sides of the neural arch remaining attached. It will be seen that the articular surface is much longer and more oblique in the former specimen. These diapophyses seem to have some points in common with those of *Manisaurus haastii*, Hector,[*] although much shorter, still this may be only the result of position in the series. At the same time it is strange that not a distinguishable fragment of a dorsal vertebra, other than these processes, has occurred with the remains of this interesting reptile.

---

* Trans. N. Zealand Inst. vi., 1874, p. 347.

Amongst the numerous fragments of ribs are several that seem to be the proximal ends of trunk ribs (Pl. vii., Figs. 5 and 6), but if so the terminal faces are cupped for the reception of the convex heads of the diapophyses already described. None are absolutely perfect, but the figures given will explain their appearance and characters almost better than words. However, the head of the rib is in each case expanded and terminally hollowed into an oval rather deep depression or cup of variable size. This appears to be similar to the structure described by Hector in *Lieodon haumuriensis.**

The proximal ends of both humeri (Pl. vii., Fig. 1) are present in the collection, and very remarkable bones they are, allied in many respects to those of *Mauisaurus haastii*, Hector.† One is three inches long, the other three and a quarter inches approximately. They are heavy and much thickened bones, the proximal articular surface hemispheric, with a sharp angular periphery and a diameter of about one and three-quarter inches. The trochanterian process is much thickened, and protuberant, and separated from the articular surface by a well marked although incomplete bicipital groove, overhung by sharp margins. The transverse diameter of the humeri is one and a half inches. The distal sloping surfaces of the bones are much roughened and pitted for muscular attachment. To one side of the trochanterian process in each is what Dr. Hector calls in *Mauisaurus* a bold rugose tuberosity "to receive the attachment of the bicipital tendon."‡ The plantar surfaces would appear to be somewhat concave.

These humeri seem to possess some important points of departure from the ordinary Plesiosaurian humerus, but are closely allied in form and character to those of *Mauisaurus*, at the same time showing sufficient differences of a distinguishing nature. For instance we see the very marked trochanterian process well separated from the articular surface, or head of the bone, by the bicipital groove, although not circumscribed by it, as such appears to be the case in *M. hastii*. We further see an equally large, although less hemispheric articular surface, and an equally strong if not stronger protuberance for the bicipital tendon, whilst the latter is somewhat differently placed to what it is in the New Zealand reptile.

Two short, transversely elongated bones are distinguishable that may be one or other of the paddle-bones, possibly the "intermedium." (Pl. vi., Fig. 6). The terminal faces are roughly facetted.

---

* Trans. N.Z. Inst., vi., 1874, p. 352.
† *Ibid*, p. 347.
‡ *Loc. cit.*, p. 348.

Phalanges are numerous throughout these remains, but I am not in a position to differentiate between them. (Pl. v., Figs. 6 and 7 ; Pl. vii., Figs. 7 – 9).

The shape of all, however, is particularly " hour-glass " like, more so than in the majority of Plesiosaurian paddles, and most of them are short, stout, and strong, with well marked articular surfaces. As the smallest seen are much constricted, longer, and have articular surfaces, the digits, like those in *Mauisaurus*, "must have been enormously prolonged to produce such attenuation."*

The only evidence of a skull consists of four teeth (Pl. vii., Figs. 10 and 11). They are small, slender, acutely conical, and curved, nearly circular in section from midway upwards, but possessing a rather more oval section below. The enamel is delicately fluted, but the teeth are not in the slightest degree carinate, as in *Pliosaurus* and *Thaumatosaurus*. None are quite perfect, but the most so, although not the largest, measures thirteen-sixteenths of an inch long by three-sixteenths in diameter at about the centre. The largest fragment, on the other hand, has a diameter of two-eighths of an inch.

The genus *Cimoliosaurus* is divided by Lydekker into two sections, the Cœlospondyline and Typical Groups. In the former the centra are excavated fore and aft, regularly amphicœlus in fact, but in the latter they are nearly flat. It follows from this that *C. leuoscopelus* belongs to the Cœlospondyli. *Mauisaurus*, Hector, with which the White Cliffs reptile agrees in more than one point, belongs to the Cœlospondyli, and had it not been so, as well as the marked difference in the vertebræ, I should have been much inclined to consider *C. leucoscopelus* as closely allied to Hector's fossil.

An epitome of our scanty knowledge of the Australian Sauropterygii will be found in the " Geology and Palæontology of Queensland."† Two species at least, perhaps three, are believed to exist, and possibly both the former are referable to *Cimoliosaurus*. They are *Plesiosaurus macrospondylus*, McCoy, and *P. sutherlandi*, McCoy. It is to be regretted that my friend, Mr. R. Lydekker, in the British Museum " Catalogue of Fossil Reptilia and Amphibia,"‡ relied on a second-hand reference to these forms, for although they have never been adequately described, still, I think they deserve a better fate than mere relegation to the limbo of MS. names.

---

* *Loc. cit.*, p. 349.

† Jack and Etheridge, Junr., 1890, pp. 508-9.

‡ Pt. 2, 1889, p. 247.

As regards *P. sutherlandi*, I have endeavoured to supply further details than those given by Sir F. McCoy from an examination of the type specimens, kindly placed at my disposal by him, supplemented by additional material from Queensland.[*] I believe *C. leioscopelus* to be distinct from both these forms. The flattened articular portions of the centra in *P. sutherlandi*, and the "more flattened than concave" centres of the cervical vertebræ appear to separate this species. In *P. macrospondylus* the edges of the articular surfaces of the centra are rugose, and thereby wholly differ from those of our form.

Of the New Zealand species, *C. australis*, Owen (+ *P. crassicostatus*, Owen),[†] possesses cervical vertebræ with flattened centra terminations, a distinct median pit in each, four hæmal foramina, and the neural arches and ribs persistently independent of the centra. The difference between this and the corresponding structure in *C. leioscopelus* is manifest. In *C. hoodii*, Owen,[‡] the hæmal surface of the cervicals is broad and flat, and there is no special transverse oblong depressions in the middle of the articular surfaces of the centra.

The cervical vertebræ of *P. holmesi*. Hector,[§] have flat terminal faces, and a humerus referred to this species, has the articular head divided by a bicipital notch, not a groove as in the present case.

P. traversi, Hector,[||] and *P. mackayi*, Hector,[**] are less known forms; the first is said to possess quadrate vertebral centra. Lastly, the absence of cervical vertebræ in the type specimens of *C. caudalis*, Hutton,[††] renders a comparison difficult.

. . . .

[*] Etheridge, Junr., Ann. Rep. Dept. Mines N.S.W., 1887 [1888], p. 167, t. 1, f. 1-4.

[†] Geol. Mag., vii., 1870, p. 51, t. 3, f. 4-5.

[‡] *Ibid.*, f. 1-3; Hector, Trans. N. Zealand Inst., vi., 1874, p. 343.

[§] Hector, *loc. cit.*, p. 344.

[||] *Ibid.*, p. 344.

[**] *Ibid.*, p. 345.

[††] *Ibid.*, xxvi., 1894, p. 354.

ON THE OCCURRENCE OF THE GENUS *COLUMNARIA* IN THE
UPPER SILURIAN ROCKS OF NEW SOUTH WALES.

By R. ETHERIDGE, JUNR., Curator.

(Plate viii.)

I BELIEVE I am correct in stating that *Columnaria* has not so far
been recognised as an Australian genus of Palæozoic Corals. When
I had the pleasure of examining the Museum at St. Stanislaus
College, Bathurst, a few months ago, under the guidance of the
Rev. Father Dowling, I observed a coral from Molong, that I
took to be *Columnaria* from macroscopic characters only, subsequently confirmed, however, by microscopic. At any rate if
the coral in question be not a species of this remarkable genus,
then the candid confession of my ignorance as to its systematic
position must be made. Father Dowling courteously allowed me
to divide the specimen, a portion of which is now in the Australian
Museum.

The composite corallum (Pl. viii., Fig. 1) is small, hemispherical,
but whether flat, rounded, or subpedunculate at the base, I am
unable to say. The colony only measures about two inches square,
and is thus even less than in *C. calicina*, Nich. The surface is
covered with shallow polygonal calices that are circumscribed by
prominent margins, crenulated by the strongly marked septa very
distinctly visible in a weathered specimen. The corallites are
closely compacted, contiguous, and completely united by their walls.
Tetragonal, quadrangular, pentagonal, hexagonal, or even irregular
corallites were observed, in contact throughout their entire course,
without any partial separation, even near the mouths as in *C.
calicina*, Nich., or some conditions of *C. alveolata*, Goldf. In
thin sections prepared for the microscope, the walls are found to
be composed of uniform grey sclerenchyma (stereoplasma), with
only here and there any trace of a primordial wall separating
them as a thin brown line ; the amalgamation is therefore so perfect that nearly all trace of primordial demarcation is practically
lost. Thus, in one instance, there is to be noted a decided departure from the microscopic structure of *Columnaria* described by
Nicholson.* The corallites have a very constant diameter of one
millimetre. In longitudinal sections (Pl. viii., Fig. 7) the same appearances are visible, the corallites also presenting the narrow tube-like
structure of the Favositidæ, but without the mural pores of the
latter. There are only sixteen septa, equally divided into primary
and secondary, the former extending across the visceral chambers

---

* Tabulate Corals Pal. Period, 1879, p. 192.

for about one-third of their diameter, or perhaps a little more, whilst the latter are mere marginal crenulations. These primary septa are tapering and spike-like in cross-section, although at the same time true lamellæ, extending from top to bottom of the corallites. There is not the faintest trace of any meeting of the septa in the calice centres to form a spurious columella as described in *C. rigida*, Billings.* At first sight the corallites appear to be provided with very few septa in consequence of the small size of the secondary, and even these are set far apart. Furthermore, the septa do not spring sharply from the corallite walls, but in consequence of the inner edges of the latter being concave between them, a more or less festoon-like appearance is given to the cross-section of each corallite, somewhat as one sees in the genus *Heliolites*. These appearances at first caused doubt to arise in my mind as to the propriety of referring this coral to *Columnaria*, but on referring to Prof. Alleyne Nicholson's excellent figures,† I found that in both *C. alveolata* and *C. calicina* very much the same features existed.

The stereoplasmic thickening of the septa is unequal, some being thin and spike-like; others, from a greater preponderance of this deposit, becoming club-shaped (Pl. viii., Fig. 6). In some corallites the secondary septa become scarcely, if at all developed, in others they assume the character previously described.

In no instance have I noticed an undue predominance in length of one or more septa, a point in which *C. pauciseptata* differs from *C. calicina* at least, but there is certainly no regular development of four septa as in *Stauria*, nor the slightest trace of a division into cycles. Many of the corallites are partially infilled with a dendritic growth of iron oxide fringing the septa.

On longitudinal weathered surfaces, the primary septa appear as strong continuous lamellæ, their paucity and larger comparative size rendering them conspicuous objects.

The tabulæ (Pl. viii., Fig. 7) are particularly well developed, simple, complete, mostly horizontal, very seldom thickened, opposite in contiguous tubes, or very slightly alternating, in other words sub-opposite. They vary from three-quarters to one millimetre apart, and in a few rare instances are somewhat more distant from one another. The non-horizontal are simply bent or curved in some portion of their course, never vesicular or incomplete. The diaphragm forming the floor of the calice is striated by the septa passing on to it. The intertabular or old visceral chambers are nearly square, from the fact of the transverse measurements of the tubes and the distance apart of the tabulæ being so nearly coincident.

There is not the slightest trace of the existence of the mural pores, or intramural canals, so characteristic of the Favositidæ.

---

* *Teste* Nicholson, *loc. cit.*, p. 196.

† *Loc. cit.*, t. 10, f. 1 and 2.

If this fossil be not a *Columnaria*, but a Favosite, then only one of two explanations is possible. Either the mural pores are confined to the angles of the prismatic tubes, or they are effaced by "complete recrystallisation or replacement." The former state could hardly exist without some trace of them being visible in one or another of the tube vertical sections, whilst the coral has not undergone sufficient alteration for the pores to be wholly effaced by the latter process. Had there been the slightest trace of these structures, I should have regarded this coral simply as an aberant form of that large and important family.

Increase took place by intra-mural gemmation, the interpolation of new tubes produced from the lip of the calicine wall of a pre-existing corallite, of which there are several instances in the longitudinal sections before me. In transverse sections these young tubes are triangular or quadrangular in outline, and situated in the angles between the older. The method of increase therefore accords with that of *C. alveolata*, Goldf., and differs from that of *C. calicina*, Nich.

The main points relied on for the identification of this coral as a *Columnaria* are (1) the absence of mural pores combined with the general Favositiform structure of the corallites, both points strongly insisted on by those who have written on this group; (2) the great regularity of the tubes and tabulæ, producing at once an entirety that is difficult to put into words, but apparent to any one who has examined authenticated examples of *Columnaria*, or as it was at one time better known, *Favistella*; (3) the absence of distinctive features of any other genus at all resembling it on a cursory examination. Under these circumstances I beg to propose for it the name of *Columnaria pauciseptata*, in allusion to the limited number of septa present, a point that will now be briefly touched on again.

Although numerous new species have, more or less perfectly, been described, indeed the late Prof. Ferdinand Roemer* recorded no less than eleven, only about three seem to be at all well known, and these chiefly through the labours of my old friend Prof. Alleyne Nicholson.† They are *C. alveolata*, Goldfuss(*non* Hall, *nec* Billings, Rominger, &c.), *C. calicina*, Nich., and *C. ? halli*, Nich. (= *C. alveolata*, Hall, Billings, Rominger, &c., *non* Goldfuss.)

In *C. alveolata* there are in all 24-30 septa, although Rominger‡ says 20-30, the primaries sometimes extending to the centre of the calices ; in *C. calicina* 28 ; in *C. ? halli* 20-40, and all quite

---

* Lethaea palæozoica, 1883, Lief. 2, p. 464.

† *Loc. cit.*. pp. 191-202.

‡ Report Geol. Survey Michigan, Lower Peninsula, iii., 1876, Pt. 2, p. 91 (as *C. stellata*).

marginal; in *C. reticulata*, Salter,[*] 36, the primaries extending half-way to the centre of the visceral chambers; in *C. franklini*, Salter,[†] although the number is not stated, they are very numerous and evidently quite marginal, like those of *C. calicina*; and in *C. gotlandica*, Ed. & H.,[‡] 36-44. The literature of those forms described by Billings is not accessible to me, and I am thus unable to enter into any comparison between his species and *C. pauciseptata*. The great dissimilarity existing between the last-named and those I have just quoted will at once be apparent, for in no instance have I observed more than sixteen septa, a disparity that can have no other than a specific significance.

As compared with the tabulæ of other species, those of *C. pauciseptata* may be said to be distant from one another. In *C. alveolata* there are three in one line, horizontal or slightly flexuous according to Nicholson, whilst Romenger says flat only; in *C. calicina* the same; in *C.? halli* the tabulæ appear to approach nearer to those of our species in distance from one another, and are horizontal and strong; in *C. reticulata* the tabulæ are "very close, four or five in the space of a line"; in *C. franklini* they are very closely packed, about four in the space of a line." Both these Arctic species, from the absence of mural pores, must be regarded as *Columnariæ*, although they have much the appearance of massive Favosites of the *F. gothlandica* group, in which the walls have undergone so much secondary alteration that the pores are not visible, a fact well known to many microactinologists. Salter's opinion, is borne out by the absence of any reference in Mr. Etheridge's description[§] of mural pores in the same corals, collected by the Nares Arctic Expedition. In *C. gothlandica* the tabulæ are said to be from one and a-half to two millimetres apart, even more distant than in *C. pauciseptata*.

The species of *Columnaria* are Silurian in their stratigraphical distribution, both Lower and Upper, with the exception of a doubtful Devonian form described by Schlüter.[‖]

The study of this coral leads me to support Prof. Alleyne Nicholson's view that *Columnaria* cannot be placed near the Favositidæ, but as suggested by Prof. Verrill, and afterwards adopted by the former, is much more nearly allied to the Astræidæ, although I have not observed in *C. pauciseptata* any trace of endothecal structures except tabulæ.

Type. In St. Stanislaus' College Museum, Bathurst.

---

[*] Sutherland's Journ. Voy. Baffin's Bay, &c., ii., 1852, p. ccxxix., t. 6, f. 2, 2a.
[†] *Ibid.*, p. ccxxix., t. 6, f. 3, 3a.
[‡] Archiv. Mus. Hist. Nat., v. 1851, p. 309.
[§] Quart. Journ. Geol. Soc., xxxiv., 1878, p. 586.
[‖] Abhandl. Geol. Specialkarte Preuss.-Thür. Staaten, viii., 1889, Heft 4, p. 14.

DESCRIPTION of TWO NEW AUSTRALIAN PHASMAS,
TOGETHER WITH A SYNOPSIS OF THE PHASMIDÆ IN AUSTRALIA.

By W. J. RAINBOW, Entomologist.

(Plates ix., x.)

The first of the two Phasmas described below is of more than ordinary interest, not only on account of its size and beauty, but also from the fact that although very close to the genus *Acrophylla*, Gray, it differs from that by the great length of its ovipositor. In *Acrophylla* the ovipositor is boat-shaped, keeled below, and does not exceed the end of the abdomen. Kirby, in a paper "On the Phasmidæ of Madagascar,"* enumerates a small collection of four previously known species, and describes a fifth, for which he founds a new genus, *Enetia*, the characters of which are :— "*Female.*—Allied to *Acrophylla*, but with the head and pro-notum spined above; wings not longer than broad; ovipositor boat-shaped, keeled below, extending considerably beyond the abdomen." In the species described below, the head and pro-notum are devoid of spines, and the wings are somewhat longer than broad, consequently it will have to come in between *Acrophylla*, Gray, and *Enetia*, W. F. Kirby.

Order ORTHOPTERA.

Family PHASMIDÆ.

Sub-Family PTEROPHASMINA.

Genus Clemacantha, *gen. nov.*

Characters of Genus.— ♀ allied to *Acrophylla*; wings longer than broad; ovipositor boat-shaped, keeled below, extending considerably beyond the abdomen.

CLEMACANTHA REGALE, *sp. nov.*

(Pl. ix., Figs. 1, 2, and 3.)

♀. Measurements (in millimeters): Length from base of antennæ to tip of abdomen, excluding ovipositor and anal styles, 177; expanse of wings, 190; length of head, 9; antennæ, 20; pronotum, 8·9; meso-notum, 26·8; meta-notum, 14·1; abdomen, 111;

---

* Ann. and Mag. Nat. Hist., Vol. viii., 6th Series, 1891, pp. 150-151.

anal styles, 12·5; ovipositor, 42; anterior femora, 36·5; hind femora, 40; tegmina, 40; greatest width of abdomen, 12·5; tegmina, 14; wing, 56.

Head pale yellow above, with broad median longitudinal bar of bright blue, sides blue above, pea-green below; face green; eyes black on narrow whitish wings; occili bright, glassy, with a reddish-brown tint; antennæ 25-jointed, slightly pubescent.

Pro-notum green above, yellowish in the median line, white laterally, yellow below with green margins; meso-notum yellow with broad median bar of bright blue, sides pea-green in front, darker behind, yellow below; the upper surface, sides, and under surface furnished with spines varying in size, those above and below are blue, and those at the sides green; the spines on the ventral surface are uniform in size, ten in number, arranged obliquely and in pairs, and surrounded with patches of blue; meta-notum yellowish in front, green behind, with a median longitudinal line of blue, sides yellow in front, green behind, the lower margins are also green, and furnished with a row of small green spines; ventral surface yellow, ornamented with four transverse bars of blue, and armed with eight spines: of these the anterior bar is the shortest and narrowest, the third and fourth is the longest and broadest; with the exception of the anterior bar, which is horizontal, all are slightly curved in a forward direction, and each bar is armed with a spine near its lateral extremity.

Abdomen long, broadest at the middle, tapering, bright green above with narrow median line of blue, under-surface pea-green; anal styles long, green; ovipositor projecting considerably beyond the abdomen, green, boat-shaped, keeled below.

Tegmina, elongate, ovate, green above, with white longitudinal bars and patches, the bars suffused with purple; underneath the anterior margin is bright red, and edged with green.

Wings: above, the costal area is bright green with white longitudinal bars suffused with purple, and the base bright red; underneath, the entire surface is bright red also; membranous portion pea-green.

Legs long, slender, with denticulated ridges, mottled with green and yellow; tibii of hind pair strongly spined on the inner side; first joint of tarsi longest, and the fifth longer than the fourth; claws long and strong.

Hab. Narrabri.

Three specimens similar to the one described, but with the meso-notum less strongly spined, are in the collection of the Australian Museum, and were taken at Wide Bay, Queensland. The specimen from which the description is written was captured by Mrs. Langhorne, Oreel Station, Narrabri. The Australian vernacular name for these insects is " Native Ladies."

B

Genus Tropidoderus, *G. Gray.*

TROPIDODERUS DECIPIENS, *sp. nov.*

(Plate x., Figs. 1, 2, 3 and 4.)

♀ Measurements (in millimeters) : Length from base of antennæ to tip of abdomen, excluding ovipositor and anal styles, 130·5 ; expanse of wings, 172 ; length of head, 7·6 ; antennæ, 24·4 ; pro-notum, 6·7 ; meso-notum, 14·3 ; meta-notum, 9·2 ; abdomen, 90·6 ; anal styles, 6·5 ; ovipositor, 27·4 ; anterior femora, 32·5; median femora, 23·8 ; hind femora, 28·5 ; tegmina, 43·6 ; greatest width of abdomen, 13·5 ; anterior femora, 3·2 ; median femora, 6·8 ; anterior femora, 8·2 ; tegmina, 18·3 ; wing, 63·1.

Head, antennæ, legs, ovipositor and anal styles, green.

Pro-notum arched, moderately granulated above and below ; meso- and meta-notum keeled in the median line and laterally ; the median keel of the former only finely serrated, but the lateral keels of both more strongly so ; these two latter are also more profusely granulated on superior and inferior surfaces than the pro-notum ; meta-notum purple laterally ; at the base of the latter there is also a median patch suffused with the same colour.

Abdomen keeled above and laterally ; superior surface and sides of a pinkish colour with the exception of the lateral keels, which are green ; inferior surface green, profusely granulated. Ovipositor boat-shaped, bright green, extending slightly beyond tip of abdomen, strongly but finely granulated, keeled below.

Tegmina elongate, ovate, keeled ; the one on the right bright green above and below, that on the left bright green on the outer half of the superior surface, including the base and tip of the inner portion, the remainder creamy white.

Wings.—Costal area of each wing purple at the base, from thence to about one-third the length, there is a pale green patch sharply rounded off at its ultimate extremity ; the remainder of the costal area above and below, bright green ; the hyaline membrane nearly colourless, or with a slight greenish hue ; veins palish pea-green.

Legs simple ; median pair hollowed out at base to receive the head ; the femora of median and hind pairs flattened out to resemble foliage, their edges strongly serrated ; meta-tarsi and tarsi brownish.

Hab. Gordon.

This beautiful insect, which so strongly similates the foliage of plants, is a typical example of Australian Phasmidæ. It will be noticed in reading the above detailed description that there is a striking difference in the colouration of the tegmina, the one on the right being entirely bright green, while that on the left has a large creamy-whitish patch. In connection with this it must be explained that when at rest the latter is always folded uppermost, and is therefore exceedingly beneficial to the insect, assisting it to elude detection by predatory foes, the whitish patch in contrast with the bright green portion having the appearance of a green leaf lighted by the sun's rays filtering through the foliage.

The specimen from which the above description was written, and which therefore forms the type of the species, was presented to the Museum by Miss Ansell, of Paddington ; in addition to this, we have in our cabinet collection a specimen taken by Mr. A. J. North, of the Australian Museum, at Ashfield, in 1895.

CATALOGUE of the DESCRIBED PHASMIDÆ of AUSTRALIA.

By W. J. RAINBOW, Entomologist.

Family PHASMIDÆ.

Genus Bacillus, *Latr.*

B. brunneus, G. R. Gray, Ent. Aust., pl. vii., fig. 3 ; Syn. Phasm., p. 21.
Burm., Handb. d. Ent. ii., 2, p. 562.
Westwood, Cat. of Orthop. Insects of the British Museum, Pt. 1, Phasmidæ, p. 12. London, 1859.
Hab. Perth, W. Australia.

B. australis, Charpentier, Orth. Descr. ♂ et ♀., pl. lvii.
Westwood, *loc. cit.*, pp. 12 and 179.
Hab. Australia.

B. dolomedes, Westwood, *loc. cit.*, p. 13, pl. v., fig. 4.
Hab. Australia.

B. peristhenes, Westwood, *loc. cit.*, p. 13, pl. vii., figs. 1, 1*a.*, ♂, pl. viii., figs. 2, 2*a.*, ♀.
Hab. Australia.

B. peridromes, Westwood, *loc. cit.*, pp. 13-14, pl. viii., figs. 2*b.*, 2*c.*
Hab. Australia.

Genus Pachymorpha, *Gray.*

P. squalida, ♀ var., *loc. cit.*, p. 15, pl. xxii., figs. 4, 4*a.*, 4*b.*
Bacillus squalidus, Hope MS.
G. R. Gray, Ent. Aust., p. 3, fig. 2 ; Syn. Phasm., p. 21
(*Pachymorpha squalida*).
Seville, H. n. Orth., p. 260.
Burm., Handb. d. Ent., ii., 2, p. 562.
Hab. Australia.

P. (?) simplicipes, Serville, H. n. Orth., p. 259.
Westwood, *loc. cit.,* p. 15.
Hab. Australia.

Genus Bacteria, *Latr.*

B. eutrachelia, Westwood, *loc. cit.,* pp. 32-33, pl. xxiv., figs. 11, 11*a.*
Hab. Perth, W. Australia.

B. cœnosa, ♀, Hope MSS.
G. R. Gray, Ent. Austr., pl. ii., fig. 2; Syn. Phasm., p. 18.
B. tenuis, ♂, Hope MSS.
Larva juvenis, *B. fragilis,* Hope MSS.
G. R. Gray, Ent. Austr., pl. vii., fig. 1 ; Syn. Phasm., p. 18.
Hab. Australia.

B. frenchi, Wood-Mason, Ann. Mag. Nat. Hist. (4) xx., p. 74.
Hab. Australia.

Genus Lonchodes, *G. R. Gray.*

L. nigropunctatus, Kirby, Trans. Lin. Soc. Lond., (2) vi., 6, pp. 453-454.
Hab. Lizard Island, Queensland.

Genus Bactridium, *Saussure.*

B. couloniamum, Saussure, Rev. et Mag. de Zool., 1868, p. 66.
Hab. Australia (Chili ?)

Genus Hyrtacus, *Stål.*

H. tuberculatus, Stål, Recensio Orthopterorum, p. 67.
Hab. Australia.

Genus Acanthoderus, *G. R. Gray.*

A. spinosus, G. R. Gray.
Westwood, *loc. cit.,* p. 48.
A. spinosus, G. R. Gray, Syn. Phasm., p. 14.
Phasma (Bacteria) spinosum, G. R. Gray, in Trans. Ent. Soc., i., 1836, p. 46 (nec. *Bacteria spinosa,* G. R. Gray, Syn. Phasm., p. 43.)
Hab. Perth, W. Australia.

Genus Eurycantha, *Boisd.*

E. australis, Montrouzier.
Westwood, *loc. cit.,* p. 65, pl. i., figs. 1, 1*a,* 1*b.* ♂, figs. 2, 2*a.* ♀
Karabidion australe, Mont., Ann. Sci. de Lyon, (2) vii., 1, p. 86.
Hab. Lord Howe Island.

Genus Anophelopis, *Westw.*

A. telesphorus, Gray.
Westwood, *loc. cit.,* pp. 69-70, pl. viii., fig. 3 ♂, figs. 7, 7*a.* ♀
Hab. Perth, W. Australia.

A. periphanes, Westwood, *loc. cit.,* p. 70, pl. viii., figs. 2, 2*a.*
Hab. Australia.

A. rhiphcus, Westwood, *loc. cit.,* pp. 70-71, pl. viii., figs. 10, 10*a.,*
10*b.*
Hab. Perth, W. Australia.

Genus Phibalosoma, *G. R. Gray.*

P. caprella, Westwood, *loc. cit.,* pp. 76-77, pl. xxi., figs. 3, 3*a.*
Hab. Australia ?

Genus Lopaphus, *Westw.*

L. gorgus, Westwood, *loc. cit.,* p. 102, pl. xi., figs. 4, 4*a.*
Hab. Richmond River, N.S.W.

Genus Xeroderus, *G. R. Gray.*

X. kirbii, G. R. Gray, Syn. Phasm., p. 32.
Burm., Handb. d. Ent., ii., 2, p. 582.
Westwood, *loc. cit.,* pp. 102-103, pl. xxxi., figs. 6, 6*a.* ♂, figs.
7, 7*a.* ♀.
Hab. Australia.

Genus Cyphocrania, *Serville.*

C. goliath, G. R. Gray.
Westwood, *loc. cit.,* pp. 107-108.
Diura goliath, G. R. Gray, Trans. Ent. Soc., i., 1836, p. 45 ;
Syn. Phasm., p. 39 *(Acrophylla G.)*
Phasma (Cyphocrania) Goliath, Audouin et Brullé, Hist.
Nat. Ins., ix., p. 105, pl. vii.
De Haan, Orth. Orient, p. 128.
Hab. Java, Timor, New Guinea, Moreton Bay, and Northern
parts of Australia.

Var. fœm. major, Cyphocrania versiruba, Serville, Orth., p.
235.
C. herculeana, Charpentier, Orth. Descr., pl. i.
Westwood, *loc. cit.,* p. 107.
Hab. Australia.

Var. fœm minor, Cyphocrania versifasciata, Serville, H. N.
Orth., p. 235.
Westwood, *loc. cit.,* pp. 107-108.
Hab. ?

C. enceladus, G. R. Gray.
  Westwood, *loc. cit.*, p. 108.
  Acrophylla enceladus, G. R. Gray, Syn. Phasm., p. 39.
  Hab. Australia.

C. pasimachus, Westwood, *loc. cit.*, pp. 109-110, pl. ix., figs. 5, 5*a*.,
    5*b*.
  Hab. Australia.

Genus Lopaphus, *Westwood*.

L. macrotegmus, Tepper, Trans. Roy. Soc. S. Austr., vol. ix., p.
    112, pl. vi.
  Hab. Mount Lofty Ranges, S. Australia.

Genus Ophicrania, *Kaup*.

O. striaticollis, Kaup, B.E.Z., p. 38.
  Hab. Australia.

Genus Acrophylla, *G. R. Gray*.

A. titan, Macleay.
  Westwood, *loc. cit.*, p. 114.
  Phasma titan, Macleay, in King's Survey of Australia, ii., p.
    454.
  G. R. Gray, Ent. Aust., i., pl. 4 ♀ *(Diura titan); ejusd.*
    Syn. Phasm. p. 39 *(Acrophylla titan)*.
  Servielle, II. N. Orth., p. 231.
  Burm., Handb. d. Ent., ii., 2, 579 *(Cyphocrania titan)*.
  Laporte, H. N. Inst.. v., pl. iv., ♂.
  Phasma (Cyphocrania) titan, De Haan, Orth. Orient., p. 129.
  Hab. Australia.

A. briareus, G. R. Gray.
  Westwood, *loc. cit.*, p. 114.
  Diura briareus, G. R. Gray, Trans. Ent. Soc., vol. i., 1836,
    p. 45 ; Syn. Phasm., p. 40.
  Hab. Australia.

A. chronus, G. R. Gray.
  Westwood, *loc. cit.*, p. 114.
  Ctenomorpha marginipennis, ♂, G. R. Gray, Ent. Austr., i.,
    pl. i., fig. 2 ; Syn. Phasm., p. 41.
  Phasma (Cyphocrania) marginipennis, De Haan, Orthop.
    Orient., p. 131.
  Diura chronus, ♀, G. R. Gray, Ent. of Austr., i., pl. v., fig.
    2 ; Syn. Phasm., p. 39 *(Acrophylla c.)*
  Servielle, H. N. Orth., p. 232.
  Burm., Handb. d. Ent., ii., 2, p. 580 *(Cyphocrania c.)*
  Hab. Australia.

A. japetus, G. R. Gray.
　Westwood, *loc. cit.*, pp. 114-115.
　Ctenomorpha spinicollis, ♂, G. R. Gray, Ent. Austr., i., pl. i.,
　　fig. 1 ; Syn. Phasm., p. 41.
　Phasma (Cyphocrania) spinicollis, De Haan, Orth. Orient,,
　　p. 131.
　Dairus japetus, ♀, G. R. Gray, Ent. Aust., i., pl. v., fig.
　　1 ; Syn. Phasm., p. 40.
　Burm., Handb. d. Ent., ii., 2, p. 580 *(Cyphocrania japetus)*.
　Hab. Melville Island, N. Territory.

A. osiris, G. R. Gray.
　Westwood, *loc. cit.*, p. 115.
　Diura osiris, G. R. Gray, Trans. Ent. Soc., i., 1836, p. 46 ;
　　Syn. Phasm., p. 40.
　Hab. Australia.

A. acheron, G. R. Gray.
　Westwood, *loc. cit.*, p. 115.
　Diura acheron, G. R. Gray, Trans. Ent. Soc., i., 1836, p. 46 ;
　　Syn. Phasm., p. 40.
　Hab. Australia.

A. macleaii, G. R. Gray.
　Westwood, *loc. cit.*, p. 115.
　Ctenomorpha macleaii, G. R. Gray, Syn. Phasm., p. 41.
　Hab. Australia.

A. tessalata, Curtiss.
　Westwood, *loc. cit.*, pp. 115-116. pl. xxxv., figs. 1, 1a., 1b. ♂,
　　figs. 2, 2a. ♀.
　Ctenomorpha tessalata, ♂, Curtiss MS.
　G. R. Gray, Syn. Phasm., p. 44.
　Hab. Moreton Bay, Queensland.

A. salmacis, Westwood, *loc. cit.*, p. 116, pl. xxxvii., figs. 2, 2a.
　Hab. Northern Australia.

A. violescens, Leach.
　Westwood, *loc. cit.*, p. 116.
　MacCoy, Prodr. Z. Vict., Melbourne, 1885, dec. viii., pl.
　　lxxix., Insects, pp. 33-34.
　Phasma violescens, ♂, Leach, Zool. Misc., i., pl. ix.
　　　G. R. Gray, Ent. Austr., pl. vi., fig. 1 *(Diura v.)* ; Syn.
　　　　Phasm., p. 40 *(Acrophylla v.)*
　　　Burm., Handb. d. Ent., ii., 2, 580 *(Cyphocrania v.)*
　Diura roseipennis, ♀, G. R. Gray, Ent. Austr., i., pl. vii.,
　　fig. 1 ; Syn. Phasm., p. 41 *(Acrophylla v.)*
　　　Burm., *loc. cit.* *(Cyphocrania v.*, ♀ *)*
　Phasma (Cyphocrania c.) roseipennis, De Haan, Orth. Orient.,
　　p. 130.

A. (Diura) virginea, Stål, Recensio Orthopterorum, Stockholm,
  1875, p. 81.
Hab. Cape York, Queensland.

Genus Clemacantha, *Rainbow.*

C. regale, Rainbow, *ante,* pp. 34-35, pl. ix., figs. 1, 2, 3.
Hab. N. S. Wales and Queensland.

Genus Vasilissa, *Kirby.*

V. walkeri, Kirby, Trans. Lin. Soc. Lond., (2) vi., 6, pp. 468-469.
Hab. Queen's Islet, N.W. Australia.

Genus Podacanthus, *G. R. Gray.*

P. typhon, G. R. Gray, Ent. Austr., i., pl. ii., fig. 1 ; Syn.
  Phasm., p. 32.
  Serville, H. N. Orth., p. 230.
  Brunn., Handb. d. Ent., ii., 2, p. 581.
  Westwood, *loc. cit.,* p. 117.
  MacCoy, Prodr. Z. Vict., dec. viii., Insects, pp. 35-36, pl.
  lxxx, figs. 1, 1*a.,* 1*b.,* 1*e.*
  Var. maris, P. unicolor (totus viridis) Charpentier, Orthop.
  Descr., pl. lvi.
  Hab. Sydney and Victoria.

P. viridi-roseus, Curtis, MS.
  G. R. Gray, Syn. Phasm., p. 43 *(Podacanthus v.)*
  Westwood, *loc. cit.,* p 117.
  Hab. In Australia, Moreton Bay.

P. Wilkinsoni, Macleay, Proc. Lin. Soc., N.S.W., vi., p. 538.
  Hab. Westmoreland, N.S.W.

Genus Necroscia, *Serville.*

N. carterus, Westwood, *loc. cit.,* p. 138, pl. xv., figs. 5, 5*a.,* 5*b.*
  Hab. Australia.

N. sarpedon, Westwood, *loc. cit.,* pp. 139-140, pl. xvi., figs. 1, 1*a.*
  ♀ ; pl. xxxii., fig. 5.
  Hab. North Australia.

N. annulipes, Curtis, MS.
  G. R. Gray, Syn. Phasm., p. 37 *(Platycrana ann.)*
  Westwood, *loc. cit.,* p. 150.
  Phasma (Necroscia) annulipes, De Haan, Orth. Orient., pp.
  118-121.
  Hab. East Indies ; Malacca ; Australia.

Genus Tropidoderus, *G. R. Gray.*

T. childrenii, G. R. Gray.
  Diura typhœus, ♂, G. R. Gray, Ent. Austr., i., pl. vi.,
    fig. 2 ; Syn. Phasm., p. 40.
  Trigonoderus childrenii, G. R. Gray, Ent. Austr., i., p. 26,
    pl. iii., fig. 1.
  Tropidoderus childrenii, G. R. Gray, Syn. Phasm., p. 31.
  Burm., Handb. d. Ent., ii., 2, p. 589.
  De Haan, Orth. Orient., p. 125.
  Westwood, *loc. cit.,* pp. 165-166.
  Hab. Australia.

T. iodomus, MacCoy, Prodr. Z. Vict., Melbourne, 1885, dec. vii.,
    Insects, pp. 33-35, pls. lxix.-lxx., figs. 2 and 3.
  Hab. Victoria.

T. rhodomus, MacCoy, *loc. cit.,* pp. 35-37, pls. lxix.-lxx., figs. 1,
    1*a.,* 1*b.,* 1*c.*
  Hab. Inglewood, Victoria.

T. decipiens, Rainbow, *ante,* pp. 36-37, pl. x., figs. 1, 2, 3, 4.
  Hab. Gordon and Ashfield, near Sydney.

Genus Lysicles, *Stål.*

L. hippolytus, Stål, OR. Ent. Belg., xx., p. 65.
  Hab. Peak Downs, Queensland.

Genus Extatosoma, *G. R. Gray.*

E. tiaratum, Macleay.
  E. hopei, G. R. Gray, Ent. Austr., i., pl. viii., fig. 1 ; Syn.
    Phasm., p. 29 *(Ectatosoma h.)* Seville, H. N. Orth., p.
    285.
  Westwood, *loc. cit.,* pp. 170-171.
  Phasma tiaratum, Macleay, in King's Survey of Australia,
    App. ii., p. 455, t. B., figs 3, 4 ♀.
  G. R. Gray, Ent. Austr., i., pl. viii., fig. 2 ; Syn. Phasm., p.
    29 *(Ectatos. t.)*
  Serville, H. N. Orth., p. 286.
  Ectatosoma tiaratum, ♂ et ♀, Burm., Handb. d. Ent., ii., 2,
    p. 576.
  De Haan, Orth. Orient., p. 110, pl. x., fig. 2 ♀.
  Hab. Australia, Tasmania, et New Guinea.

E. bufonium, Westwood, Thes. Ent. Oxon., p. 174, pl. xxxii., fig. 2.
  Hab. Australia.

Genus Cladoxerus, *Kaup.*
C. insignis, Kaup, B.E.Z., p. 39.
  Hab. Australia.

Genus Ctenomorpha, *Gray.*

C. nigro-varia, Stål, Reccnsio Orthopterorum, p. 83.

Hab. Cape York, Queensland.

Genus Vetilia, *Stål.*

V. eurymedon, Stål, C.R. Ent. Belg., xx., p. lxiii.

Hab. Cape York, Rockhampton, Queensland.

DESCRIPTIONS of NEW LAND SHELLS.

By C. HEDLEY, Conchologist.

(Plate xi.)

PUPISOMA CIRCUMLITUM, *n. sp.*

(Plate xi., figs. 1, 2, 3.)

Shell globose conical, perforate, thin, translucent. Colour an uniform pale tawny olive. Whorls three and a-half, well rounded; suture impressed. Sculpture,—everywhere the whorls are crossed by fine, close, raised hair lines; at irregular intervals these tend to rise into lamellæ, which latter can scarcely be detected in profile on the periphery; the embryonic shell is similarly sculptured, no trace of spiral sculpture can be seen; a break at the completion of the second whorl suggests that here ends the nepionic shell. Umbilicus minute, funnel shaped, showing only the preceding whorl. Aperture very oblique, ovate lunate, lip simple, columellar margin broadly reflexed over the umbilicus; callus thin, transparent. Height, 1·9; breadth, 2 mm.

Type.—Australian Museum C. 3459.

Hab.—Received through Mr. C. E. Beddome from Dr. May, who gathered it on trees at Bundaberg, Queensland; also collected on orange trees near Grafton, N.S.W., by myself.

This snail conceals itself by plastering the shell over with grains of earth, etc., entangled in mucus. The device reminded me of the European *Balea perversa*, which adopts the same habit in similar situations. Occasional abrasions seem to show that the colour resides in a very thin epidermis.

I have not the advantage of being autoptically acquainted with any of the known *Pupisoma*, but the novelty corresponds so closely

to the drawings of several that I have some confidence in introducing it under that genus.   If this classification be correct, the
range of the genus is now by a leap of three thousand miles extended from Borneo and the Philippines to New South Wales;
thus introducing into Australia a fresh component of that faunal
element which Prof. Spencer has termed "Torresian."[*]

ENDODONTA WATERHOUSIÆ, n. sp.

(Pl. xi., figs. 7, 8, 9, 13, 14.)

Shell sub-discoidal, spire sunk, widely umbilicated, opaque, dull.
Colour, on a ground of pale buff above irregularly splashed with
madder brown which beneath tends to flow in irregular, oblique
and zig-zag lines, apex pale straw.   Whorls four and a half, rounded
except for a flattening between the suture and the periphery, the
first three whorls slightly and gradually ascending above the apex,
the last half whorl broadening slightly and gradually and slightly
descending.   Sculpture consisting of sharp lamellate ribs, slightly
flexed at their origin at the suture, then crossing the whorl at
right angles, curving backwards and downwards to the periphery,
thence taking a straight course to the lip of the umbilical crater,
over the edge of which they curve forwards ; on the last whorl
these ribs number ninety-five, on the penultimate fifty-three, and
on the antipenultimate thirty-two ; they crowd closer as the whorls
proceed, but the spacing is not always uniform ; on the last whorl
the interstices are as broad or twice as broad as the ribs, on the
final sixth, however, the ribs tend to obsolescence ; most minute
hair lines, parallel to the major sculpture, occupy these interstices ;
the ribs cease entirely and suddenly at the initial whorl and a-half,
which by transmitted light are shown to possess radial hair-lines
decussated by equally fine spiral striæ.   Umbilicus a third of the
base of the shell in diameter, cup-shaped, exposing every preceding
whorl, coloured and sculptured like the spire.   Aperture slightly
oblique, subrhomboidal, peristome sharp, straight, even at the
columellar margin ; viewed from above the peristome describes a
wide convex, then a sharper concave curve on approaching the
insertion.   Projecting callus on body whorl steel purple, burying
the costæ in its advance.   Major diameter 7, minor 6, height 3½
mm.

Type.—Australian Museum C. 3458.

Hab.—Mount Gower, Lord Howe Island.

A specimen dissected was not in a satisfactory state for examination, and I was only able to unravel the basal portion of the
genitalia (Fig. 13).   This showed a greatly dilated reniform penis
sac on a long stalk surmounted by an equally long epiphallus ;

---

[*] Rep. Horn Expl. Exp. i., 1896, p. 197.

through the coil of the penis the tentacle is retracted. Between
the lower end of the uterus and the entrance of the spermatheca
duct a bulb occurs like the swelling in a similar situation of
certain *Trochomorpha.* The jaw (Fig. 14) is crescentic, with a
median projection, and is closely transversely striated. The
radula is formed like that of *E. coma,* Gray, as figured by
Pilsbry,[*] it consists of one hundred and thirty rows of 13:8:1:8:13.

A near ally of this very distinct species is *E. coma,* Gray, of
New Zealand, from which the Lord Howe Island form is separable
by its concave spire, closer ribbing, and larger size. *E. pinicola,*
Pfeiffer, from New Caledonia is also allied, but that has an elevated
spire, weaker ribs, and is larger. This trio of kindred *Endodonta*
supports a trio of equally related *Placostyli,*[†] viz., *P. bovinus, P.
bivaricosus,* and *P. caledonicus,* in linking together the faunas of
these islands.

This species is that recorded in my article on "The Land and
Fresh-water Shells of Lord Howe Island,"[‡] as *Charopa textrix,*
Pfr., this being the identification of Mr. J. Brazier in the Memoir
on Lord Howe Island. My suspicion of this determination was
aroused by comparing the shell with the excellent figures of
Pfeiffer,[§] but I was over-ruled by the weight of Mr. Brazier's
authority.

Under the name of *C. textrix* this shell has been widely distributed. One of the recipients, Mr. John Ponsonby, of London,
on comparing this with authentic *C. textrix* in the British Museum
found it to be a different species. Not only am I indebted to him
for this information, but he has also generously waived in my
favour his right of describing it.

Now arises naturally the question what *C. textrix* really is. I
am tempted to believe it identical with the small form of *Endodonta costulifera,* Pfr. My reasons are, that a shell from Noumea
which I identify as such closely corresponds with Pfeiffer's account
of *textrix,* that Macgillivray collected *costulifera* as well as *textrix,*
and finally that *textrix* is unknown from Lord Howe Island
which has been thoroughly searched for it. If this be so, then
the name *Endodonta textrix* must pass into synonomy.

The novelty is dedicated to Mrs. J. G. Waterhouse, an enthusiastic and accomplished conchologist of Sydney, whose assistance
in studying this and other forms I gratefully acknowledge.

---

[*] Tryon & Pilsbry—Manual Conchology (2) ix., pl. ix., f. 23.

[†] Etheridge : "A much thickened variety of *Bulimus bivaricosus* from
Lord Howe Island."—Rec. Aust. Mus. i., 1891, p. 130.

[‡] Rec. Aust. Mus., i., 1891, p. 137.

[§] Conchylien Cabinet (2) Helicea, pl. clxii., ff. 14-17.

FLAMMULINA ABDITA, *n. sp.*

(Pl. xi., figs. 10, 11, 12.)

Shell very small, thin, translucent, moderately umbilicated, and depressed. Colour raw umber, paler on the earlier whorls and purplish on the apex. Whorls.three and a half, rather rapidly increasing, rounded, channelled at the suture and slightly descending at the aperture. Sculpture,—the whorls are crossed at irregular intervals by numerous lamellate ribs, rising on the periphery into thin recurved plates but obsolete on the last quarter whorl; between and parallel to these ribs are fine raised hair-lines, which are cut by close, fine, faint, irregular spiral incised lines. At a whorl and a half the limit of the embryonic shell is sharply indicated by the commencement of the above described sculpture, the earliest whorls being smooth except for close, fine, incised, spiral lines ; on the apex is a small pit. Umbilicus about a quarter of the shell's diameter, exposing the earlier whorls. Aperture ovate lunate, slightly oblique ; peristome sharp, straight, except where reflected on the columella margin, no visible callus on the inner side. Major diameter, 1·5 ; minor, 1·3 ; height ·8 mm.

Type.—Queensland Museum.

Hab.—Collected by Mr. A. Giulianetti, in October, 1896, at a height of 12,200 feet on Mount Scratchley, British New Guinea.

This species possesses close affinities to the wide-spread Australian *(H.) paradoxa*, Cox, from which the novelty differs by its less developed ribs, less elevated spire, wider umbilicus, and smaller size. Pilsbry's figure of *Endodonta acanthinula*, Crosse, suggests to me that that New Caledonian species should be grouped herewith. The few whorls and the aspect of the embryonic shell induce me to place this Papuan atom in *Flammulina* rather than in *Endodonta*, but with our present imperfect knowledge of these groups such classification can only be considered provisional.

SITALA ? SUBLIMIS, *n. sp.*

(Plate xi., fig. 4, 5, 6.)

Shell small, thin, translucent, depressedly turbinate, narrowly perforate. Colour tawny olive. Whorls three and a half, gradually increasing in diameter, rounded. Suture impressed. Sculpture,— the otherwise smooth shell is everywhere crossed by extremely fine, close, transverse hair-lines, more prominent above, almost effaced beneath. Umbilicus extremely narrow, elliptical, exposing only the previous whorl. Aperture roundly lunate, not descending, oblique ; lip sharp, straight, except a slight reflection at the columella ; callus on body whorl thin, deposited in transverse streaks. Major diameter, 2·4 ; minor, 2 ; height, 1·5 mm.

Type.—Queensland Museum.

Hab.—Collected by Mr. A. Guillianetti, on Mt. Scratchley with the preceding and with *Rhytida globosa*, which latter has been only previously recorded from Mt. Victoria.

This shell, of which my single example may not be adult, appears allied to *(Zonites) subfulvus*, Gassies, from New Caledonia, and to *Conulus paramattensis*, Cox, from N.S.W., but is more depressed than either, smaller, and with a whorl less. A near ally is *C. starkei*, Brazier, which however has much coarser sculpture, and has also more whorls and larger size, two differences which perhaps balance one another. It belongs to a group which Tryon and others rank under *Conulus*. Thinking it improbable that this Palæartic genus extends so far, I prefer to temporarily locate it in *Sitala*, to some species of which it bears a likeness, and which has in another form been shown to reach New Guinea. Attention to kindred small forms and to the literature devoted to their elucidation (!) shows how much definitions are required for *Conulus, Microcystis, Trochonanina*, and other groups.

While on the subject of land shells from British New Guinea, it may be mentioned that the unfiguered *Succinea strubelli*, Kobelt, and *S. papuana*, Strubell,* from Cloudy Mountains and Lorne Range, North of Orangerie Bay, B.N.G., are most probably synonymous with *Succinea simplex*, Pfeiffer. In stating that his is the first record of the genus in Papua, Herr Strubell shows that he has failed to compare his supposed new species with one already illustrated and identified from New Guinea.†

Another addition to the known fauna of British New Guinea has lately come to light in specimens of *Atopos prismatica*, Tapparone, Canefri, now in the Australian Museum, collected on the Fly River, by W. W. Froggatt, during the Expedition of the Geographical Society of Australasia in 1885. It was first recorded from Sorong I., Dutch New Guinea, and then from an island of Torres Straits, Q., and the Huon Gulf, German New Guinea.‡

---

* Nachr. dent. Malak. Gesell, Oct. 1895, p. 152.

† Proc. Linn. Soc. N.S.W. (2), vii., 1892, pp. 100, 691-2, pls. xii., f. 32; xlii., ff. 34-37.

‡ Simroth, Zeits. Wiss. Zool., lii., 1891, p. 594.

DESCRIPTION OF A NEW SPECIES OF *COLLYRIOCINCLA* FROM QUEENSLAND.

BY ALFRED J. NORTH, C.M.Z.S.,
Ornithologist to the Australian Museum.

COLLYRIOCINCLA CERVINIVENTRIS, *sp. nov.*

Adult male.—General colour above greyish-brown very slightly shaded with olive, clearer grey on the head; wing-coverts like the back, the quills brown washed with olive on their outer webs, and externally edged with grey; tail brown, the two centre feathers, and the outer webs of the remainder, shaded with grey; feathers in front of the eye dull white; ear-coverts pale brown with narrow white shaft streaks; cheeks and throat white, slightly tinged with buff; remainder of the under surface and under wing-coverts pale fawn colour, the feathers on the chest shaded with grey; bill and legs fleshy-brown. Total length 7·2 inches; wing 3·7, tail 3·2, bill 0·87, depth at nostril 0·25, tarsus 1·02.

Hab.—Dawson River, Queensland.

Type.—In the Australian Museum, Sydney.

Obs.—Another specimen, probably a female has the primary-coverts and outer webs of the secondaries washed with rufous-buff. This is the inland representative of *C. rufigaster* of the coastal brushes, from which it may be distinguished by its longer and thinner bill, and by its very much paler upper and under surface. In all the specimens from this district—three in number—the distinctive characters of this species are constant.

Eggs of this closely allied species were described by me in the "Nests and Eggs of Australian Birds," * under the name of *C. rufigaster.*

The type of *C. rufigaster* was obtained in the brushes of the Clarence River, New South Wales. An adult male from this locality measures :—total length 7·3 inches; wing 3·9, tail 3·3, bill 0·8, depth at nostril 0·3, tarsus 1·08. A large series of specimens from the coastal districts of Queensland, as far north as Cairns, vary only in wing measurement from 3·8 to 3·9 inches.

---

* Austr. Mus. Cat. xii., 1889. Nests and Eggs, p. 83.

Gould's *C. parvissima* is a decidedly smaller race, and is further-more distinguished by its upper parts being more strongly washed with olive. A specimen from Cape York measures, total length 6·3 inches, wing 3·5, tail 2·7, bill 0·75. It ranges as far south as the Herbert River. The wing-measurement varies from 3·5 to 3·6 inches. In a large series of specimens examined I can find no gradation in size between *C. rufigaster* and its smaller northern ally, *C. parvissima*. Dr. Sharpe, in the "Catalogue of Birds in the British Museum,"* has described a specimen of the latter race from Cape York under the name of the larger species, *Pinarolestes rufigaster*, Gould.

On *STICHOPUS MOLLIS*, HUTTON.

By THOMAS WHITELEGGE, Zoologist, Australian Museum.

DURING a recent visit to Eden, Dr. J. C. Cox obtained a *Holothurian*, which he presented to the Museum collection.

The species proves to be *Stichopus mollis*, Hutton, hitherto only recorded from New Zealand. At first I referred it to *S. sordidus*, Theel,† but on consulting a paper by Prof. A. Dendy‡ since received, I find that the "Challenger" specimens are considered to be examples of Hutton's *Holothuria mollis*. I agree with the opinion as expressed by Prof. Dendy. There appears to be no character to distinguish them except that of colour, which is evidently variable.

The colour of the Eden example in formol is light yellowish-brown, with the disks of the pedicels and the tips of the dorsal papillæ darker.

The large bilateral plates form a ring on the margin of the disk and encircle the large central perforate plate at the extremity of the pedicel.

The dorsal papillæ are supported by curved, smooth, spiny, or branched rods, disposed transversely ; the lower ones tend to form plates similar to those of the ventral pedicels ; apically each papilla terminates in a large perforate plate, which is surrounded by a series of stout moniliform rods, with either simple or spinose ends.

Length of specimen 130 mm.

---

* Brit. Mus. Cat. Birds iii. 1877, p. 296.

† Chall. Rep. Zool., xiv., p. 162, pl. viii., fig. 3.

‡ Jour. Linn. Soc. Zool. xxvi., 1897, p. 46, pl. vii., figs. 73-82.

5th August, 1897.

The NOCOLECHE METEORITE,

With CATALOGUE and BIBLIOGRAPHY of AUSTRALIAN METEORITES.

By T. Cooksey, Ph.D., B.Sc., Mineralogist.

(Plates xii., xiii., xiv.)

The iron, which has been named the "Nocoleche" Meteorite, was presented to the Trustees of this Museum by Mr. George Raffel, in October, 1896, from whom the information was gleaned that it was found lying upon the surface of stony ground at a spot five miles south-west of Nocoleche Station, near Wanaaring, N.S. Wales. The specimen received was the whole of the mass found. Its existence was known twelve or eighteen months previously, but there is no record of any meteorite or meteoric showers having occurred in the district. The total mass weighed 20,040 grams (equal to 44·18lbs. avoirdupois). Its external form is of a pronouncedly rugged character, and the iron is raised into points and ridges, the latter trending mostly in a uniform direction. This character is shown in Pl. xii. At B (Fig. 1) is a projecting rugged nob, connected to the main mass by a neck which is almost penetrated at one point by a deep hole, very probably at one time containing a nodule of troilite. A similar hole, but considerably smaller, is situated in the large cavity at the opposite side. (Pl. xii., Fig. 2). The remains of a black magnetic coating are found in many places, where it is mostly thin, but in protected positions, patches remain which in places have a thickness of 2·5 mm. The external appearance, on arrival, however, was rusty and up to a certain level the colour was fresher than that above, suggesting that the iron was partially buried at the time of its removal. The form of the mass is no doubt partly due to weathering. The length from A to B (Plate xii., Fig 1) is twelve and three-quarter inches, from C to D eleven and a half inches, and greatest thickness, leaving out of account the projecting nob, five and a half inches. The specific gravity was found to vary slightly from place to place. One piece of the iron weighing 5·5824 grams, and visibly free from troilite, had a specific gravity of 7·721 (uncorr.); while another piece weighing 2·2798 grams, had a specific gravity of 7·796 (uncorr.) The specific gravity of a large piece weighing just over seventeen ounces, and containing small nodules of troilite (apparently a fair sample of the whole mass) was taken at the Royal Mint, Sydney, and found to be 7·69.

The meteorite was cut by Prof. H. A. Ward, of Rochester, U.S.A., and the surface shown in Plates xiii. and xiv., etched to

A

within a quarter of an inch from the edge. The plates are about
two-thirds of the natural size and represent the Widmanstätten
figures as seen under different aspects of reflected light. The
markings consist of bands of beam iron (kamacite), running in
three directions, which cross each other approximately at angles
of 60°. Under the glass the bands themselves show usually two,
and sometimes three series of finely-etched parallel lines, crossing
at varying angles. Troilite is freely distributed throughout,
occurring in nodules (Pl. xiii., right-hand top corner), and in
the smaller patches and cracks. The latter are numerous, and
mostly separate the bands of beam iron from each other. The largest
nodule observed was one and a half cmm. long and one cm. wide, and
possessed a dark bronze-like metallic lustre. The nodules are lined
by a darker substance, usually forming a very thin layer, which is
thickened in places and continued into the cracks. The iron immediately surrounding the nodules is somewhat more brilliant
than that further removed, but no defined line is generally to be
seen separating this brighter iron from the remainder. The etched
iron shews in places a very fine irregular mottling, forming occasionally more or less regular lines ; but this formation appears to
be independent of the crystalline structure. Very small specks
and strings of bright particles are very sparingly distributed
throughout the iron, and in a relatively larger quantity occur with
the troilite. By dissolving 6·2114 grms. of the iron in hydrochloric
acid in the cold, a residue containing ·0386 grm. of a black
powder and ·0014 grm. of bright metallic particles was left undis-
solved. The latter under the microscope were seen to consist of a
mixture of brilliant grains and needles of a steel grey colour. The
mass of the iron is almost entirely a mixture of beam iron (kamacite) and troilite, but taenite and plessite do not appear to be
developed. The bright grains and prisms are no doubt a mixture
of the phosphides of iron and nickel (schreibersite and rhabdite),
and the black powder of carbon and carbide of iron. I hope,
however, to have a further opportunity of more closely examining
this residue. The small quantity of residue insoluble in boiling
acids consists mainly of carbon.

Analysis.—To obtain an average sample for analysis, about
twenty grams of small chippings were cut off from portions visibly
free from troilite.

I. 3·0702 grams of the above were dissolved in hydrochloric
acid, and after separating the residue and precipitating the trace
of copper present, the method of analysis was that adopted by
Stanislas Meunier.* From the solution acidified with acetic acid,
the nickel, cobalt and part of the iron were precipitated by

---

* Stanislas Meunier, Encyclopédie chimique, ii., 1884, App. 2, Météorites, p. 26 *et seq.*

sulphuretted hydrogen. The remaining portion of iron was then thrown down by making the solution alkaline with ammonia.

II. 3·0086 grams were separately dissolved in a mixture of nitric and hydrochloric acids to estimate the sulphur and phosphorus.

III. 2·5045 grams were dissolved in hydrochloric acid, the residue and copper separated, the filtrate made up to 500 cc., and 50 cc. of this taken to estimate the iron, nickel, and cobalt. The iron was separated from the nickel by precipitation with ammonia.

IV. A small piece of very bright iron, shewing brilliant cleavage surfaces, and weighing ·3733 gram, was analysed for iron. It was obtained from the vicinity of a troilite nodule.

	I.	II.	III.	IV.
Residue	·09	...	·07	...
Copper	·07	...	·03	...
Iron	97·09	...	96·65	97·05
Nickel	2·91	...	} 3·69	...
Cobalt	·21	...		...
Phosphorus ...		·12	...	...
Sulphur		·11	...	...
	100·37		100·44	

The iron is active towards acids and sulphate of copper.

TROILITE.

The material for analysis was obtained from a nodule, the specific gravities of two pieces of which, weighing ·3746 gram, and 1·3704 grams, were found to be respectively 5·50 and 5·442. On powdering and treating with a concentrated solution of sulphate of copper, a copious precipitate of metallic copper was very quickly formed. To remove all the iron the powder was boiled with a concentrated solution of sulphate of copper for a quarter of an hour.* A black product was finally obtained after washing, which had a specific gravity of 4·66, and was found to be sulphide of copper containing only 6·94 per cent. of iron. An attempt was then made to separate the troilite from impurities by repeatedly washing and separating the lighter and heavier portions. A product was in this manner obtained which curiously had a specific gravity of 4·788, but on analysis was found to contain 20·32 per cent. of sulphur and 73·49 per cent. of iron. The iron must therefore be intimately mixed with the troilite. It was finally ascertained that the iron could be removed by standing for some hours with a solution of either sulphate of copper or

---

* Stanislas Meunier, Encyclopédie chimique, ii., 1884, App. 2, Météorites, p. 57 ; W. Crookes, Select Methods in Chemical Analysis, 1886, p. 201.

chloride of mercury in the cold, the sulphide of iron being unacted upon under these circumstances. The latter method of purification was chosen. ·18 gram of iron was extracted from ·52 gram of the original powdered sulphide by allowing the latter to stand for twenty-four hours with a concentrated solution of mercuric chloride in the cold. This is equal to 34·6 per cent.

The purified troilite had a specific gravity of 4·645 (uncorr.), and an analysis of ·2408 gram. gave :—

Residue	trace.
Iron	62·01
Nickel and Cobalt...	·89
Sulphur	38·28
	101·38

From the analysis of the iron it will be noticed that the quantity of nickel present is unusually small. It is not exceptionally so, for several irons have been previously analysed containing about a similar or even smaller quantity.

An examination of the Widmanstätten figures shews the iron to be an octahedrite whose width of lamellæ vary from 1mm. to 2·5mm., the greater number, however, lying between the limits of 1mm. and 2mm. Following Dr. A. Brezina[*] in his provisional system of classification, it would be placed in Group 47, containing octahedrites with broad lamellæ (symbol Og) and therefore classed with the Cranbourne (Victoria) and Youndegin (W. Australia) meteorites more closely, and with the Cowra (N.S. Wales), Mooibi (N. S. Wales), Temora (N. S. Wales), Mungindi (Queensland), and Thunda (Queensland) meteorites as being with them an iron having an octahedral crystalline structure. The Nocoleche iron, however, agrees much more closely with the Murfreesboro one, both as regards crystalline structure and relative proportions of iron and nickel. The two figures given by Dr. A. Brezina[†] of the latter would very well represent the structure of the former, there being merely a slight difference in the average width of the lamellæ.

The percentages of iron and nickel in the Murfreesboro iron given by de Troost,[‡] are as follows :—

Iron	96·0
Nickel	2·4
Residue	1·6
	100·0

[*] Ann. K. K. Naturhist. Hofmus. Wien., x., 3–4, 1895, p. 85.
[†] Ibid, p. 276.
[‡] Silliman's Amer. Journ. Sci. (2), v., p. 351, and ibid., xv., p. 6.

CATALOGUE OF AUSTRALIAN METEORITES.

The following Catalogue represents all the Australian Meteorites discovered up to date, so far as known to me. The name of the meteorite is first given, followed by the more important references

ADDENDUM.

SINCE going to press, the following additional information has been received in continuation of "Catalogue of Australian Meteorites," pages 55 – 60 :—

YARDEA.—

Type.—Siderite.

Weight.—7 lbs. 3½ozs.

Locality.—Four miles S. of Yardea Station, Gawler Ranges, South Australia.

Finder and Date.—Found in 1875.

Coll.—Public Museum, South Australia.

Type.—Siderolite.

Weight.—48lbs.

Loc.—Near Baratta, No. 1.

Finder and Date.—

Coll.—H. C. Russell, F.R.S., Govt. Astronomer, Sydney.

chloride of mercury in the cold, the sulphide of iron being unacted upon under these circumstances. The latter method of purification was chosen. ·18 gram of iron was extracted from ·52 gram of the original powdered sulphide by allowing the latter to stand for twenty-four hours with a concentrated solution of mercuric ·· ·· ·· ·· ·· ·· This is equal to 34·6 per cent.

100·0

---

* Ann. K. K. Naturhist. Hofmus. Wien., x., 3 - 4, 1895, p. 85.
† Ibid, p. 276.
‡ Silliman's Amer. Journ. Sci. (2), v., p. 351, and ibid., xv., p. 6.

CATALOGUE OF AUSTRALIAN METEORITES.

The following Catalogue represents all the Australian Meteorites discovered up to date, so far as known to me. The name of the meteorite is first given, followed by the more important references bearing on it. These are succeeded by the type, locality, finder and date, and the collection in which the stones, or slices, or both, are to be seen. Every effort has been made to obtain accurate details, but gaps in the information must of necessity occur when objects like these, often passing through several hands before coming under the eye of the describer, are being dealt with.

BALLINOO.—H. A. Ward, Supplementary Catalogue of Meteorites, April 1, 1897.
Type.—Siderite ?
Weight.—92lbs.
Loc.—Ten miles S. of Ballinoo, Murchison River, W. Australia.
Finder and Date.—George Denmack, in 1892.
Coll.—H. A. Ward, Rochester, U.S.A.

BARATTA, No. 1.—A. Liversidge, Trans. Roy. Soc. N.S. Wales, 1872, p. 97 ; *Ibid.*, xiv., for 1880, p. 308 ; *Ibid.*, xvi., for 1882 (1883), p. 31 ; A. Brezina, Ann. K.K. Naturhist. Hofmus. Wien., x., 3–4, 1895, p. 252.
Type.—Siderolite, classed with the Black Chondrites.
Weight.—145lbs.
Loc.—Baratta Station, thirty-five miles N.W. of Deniliquin, N.S. Wales.
Finder and Date.—F. Gwyne, of Murgah, 1852.
Coll.—H. C. Russell, F.R.S., Govt. Astronomer, Sydney.

BARATTA, No. 2.—A. Liversidge, Proc. Austr. Assoc. Adv. Sci., ii., for 1890 (1891), p. 387 ; H. C. Russell, Journ. Roy. Soc. N.S. Wales, xxiii., 1889, p. 46.
Type.—Siderolite.
Weight.—31lbs.
Loc.—Near Baratta, No. 1.
Finder and Date.—
Coll.—H. C. Russell, F.R.S., Govt. Astronomer, Sydney.

BARATTA, No. 3.—A. Liversidge, Proc. Austr. Assoc. Adv. Sci., ii., for 1890 (1891), p. 387 ; H. C. Russell, Journ. Roy. Soc. N.S. Wales, xxiii., 1889, p. 46.
Type.—Siderolite.
Weight.—48lbs.
Loc.—Near Baratta, No. 1.
Finder and Date.—
Coll.—H. C. Russell, F.R.S., Govt. Astronomer, Sydney.

BINGARA.—A. Liversidge, Journ. Roy. Soc., xiv., 1880, p. 308 ; *Ibid.*, xvi., for 1882 (1883), p. 35 ; A. Brezina, Ann. K.K. Naturhist. Hofmus. Wien., x., 3 - 4, 1895, p. 294.

Type. -Siderite, belonging to the Hexahedrite Group.

Weight.—240·7 grams.

Loc.—Bingara, N.S. Wales.

Finder and Date. -1880.

Coll.— Greater portion in the Mining and Geological Museum, Sydney ; slice at the Hofmuseum, Vienna ; A. Liversidge, F.R.S., University, Sydney.

COWRA.- G. W. Card, Rec. Geol. Survey N. S. Wales, v., 2, p. 51 ; A. Brezina, Ann. K.K. Naturhist. Hofmus. Wien., x., 3 - 4, 1895, p. 267.

Type.—Siderite, belonging to the Octahedrite Group.

Weight.—12¼lbs.

Loc.—Summit of Battery Mountain, junction of the Burrowa and Lachlan Rivers, near Cowra, N.S. Wales.

Finder and Date.—Mr. John O'Shaughnessy, before 1888.

Coll.--Mining and Geological Museum, Sydney ; small slices at the British Museum (Nat. Hist.), and Hofmuseum, Vienna.

CRANBOURNE (OR BRUCE), No. 1.-- W. Von. Haidinger, Sitzungsber. Akad. Wiss. Wien., xliii., 1861, p. 583 ; *id.*, 1861, xliv., pp. 378 and 465 ; *id.*, xlv., 1862, p. 63 *(fide* Flight) ; W. Flight, Phil. Trans., clxxiii., 1882, p. 885 ; G. Foord, Brough Smyth's Gold Fields and Mineral Districts of Victoria, 1869, p. 424 ; K. Hauskofer, Journ. Prakt. Chem., cvii., 1869, p. 333 ; M. Berthelot, Ann. Chim. et Phys. xxx., 1873, p. 419 ; A. Brezina, Ann. K.K. Naturhist. Hofmus. Wien., x., 3 - 4, 1895, p. 285.

Type. -Siderite, belonging to the Octahedrite Group.

Weight. -3 - 4 tons.

Loc.—Cranbourne, near Western Port, Victoria, Lat. 38° 11' S., Long. 145° 20' E.

Finder and Date.—Existence known in 1854.

Coll. —British Museum (Nat. Hist.).

CRANBOURNE, No. 2.— W. von Haidinger, Sitzungsber. Akad. Wiss. Wien., xliv.. 1861, pp. 378 and 465 ; *id.*, xlv., 1862, p. 63 ; W. Flight, Phil. Trans., clxxiii., 1882, p. 885.

Type.—Siderite.

Weight.—Several hundredweights.

Loc.—Near Cranbourne, 3·6 miles N. of Cranbourne, No. 1, in Lat. 38° 8′ S., Long. 145° 22′ E.

Finder and Date.—Existence known in 1854.

Coll.—Technological Museum, Melbourne.

ELI ELWAH.—A. Liversidge, Proc. Austr. Assoc. Adv. Sci., ii., for 1890 (1891), p. 388.

Type.—Siderolite.

Weight.—33½ lbs.

Loc.—Eli Elwah Station, fifteen miles W. of Hay, N.S. Wales.

Finder and Date.—

Coll.—H. C. Russell, F.R.S., Govt. Astronomer, Sydney.

GILGOIN, No. 1.—H. C. Russell, Journ. Roy. Soc. N.S. Wales, xxiii., 1889, p. 47 ; A. Liversidge, Proc. Austr. Assoc. Adv. Sci., ii., for 1890 (1891), p. 388.

Type.—Siderolite.

Weight.—67 lbs. 5 ozs.

Loc.—Gilgoin Station, forty miles E.S.E. of Brewarrina, N.S. Wales.

Coll.—H. C. Russell, F.R.S., Govt. Astronomer, Sydney.

GILGOIN, No. 2.— H. C. Russell, Journ. Roy. Soc. N.S. Wales, xxvii., 1893, p. 361.

Type.—Siderolite.

Weight.—74 lbs. 5 ozs.   Most probably part of the same meteorite as Gilgoin, No. 1.

Loc.—Two miles S. of Gilgoin, No. 1.

Coll.—H. C. Russell, F.R.S., Govt. Astronomer, Sydney.

HADDON.—*Illustrated Australian News*, 17th May, 1875, p. 68 ; W. Flight, Geol. Mag., (2), ix., 1882, p. 107.

Obs.—A meteor was seen on April 14th, at 0·30 a.m., and immediately afterwards an eyewitness thought he saw matter fall near him. Several pieces of melted matter of varying colour were found.

Type.—Aerolite ?

Loc.—Haddon, Grenville Co., Vict.

Coll.—

LE GOULD METEORITE.—Le Gould, Geol. Mag., i., 1864, p. 142.

Obs.—An aerolite was found ten inches in diameter, which had struck and broken a tree.

Loc.—Two day's march beyond the Isaacs, the first branch of the Mackenzie River, Queensland.

MOONBI.—J. C. H. Mingaye, Journ. Roy. Soc. N.S. Wales,
        xxvii., for 1893 (1894), p. 82 ; A. Brezina, Ann. K.K.
        Naturhist. Hofmus. Wien.. x., 3 - 4, 1895, p. 268.
Type.—Siderite, belonging to the Octahedrite Group.
Weight.—29lbs.
Loc.—Top of one of the ridges of the Moonbi Range, eighteen
        miles from Moonbi Township, N.S. Wales.
Finder and Date.—Mr. Langston, in 1892.
Coll.—Technological Museum, Sydney (main mass).

MOORANOPPIN.—H. A. Ward, Supplementary Catalogue of Meteor-
        ites for sale, April 1, 1897.
Type.—Siderite?
Weight.—
Loc.—Mooranoppin, West Australia.
Finder and Date.—An Aboriginal in or before 1893.
Coll.—Perth Museum, Perth, West Australia ; H. A. Ward,
        Rochester, U.S.A.

MOUNT STIRLING.—(Under investigation).
Type.—Siderite.
Weight.—200¼lbs.
Loc.—Twenty-five miles S.E. of Mount Stirling, one hundred
        and thirty miles E. of Perth, West Australia.
Finder and Date.—Existence known in 1892.
Coll.—Australian Museum, Sydney.

MUNGINDI, No. 1.—G. W. Card, Rec. Geol. Surv. N.S. Wales,
        v., 3, 1897, p. 121.
Type.—Siderite, belonging to the Octahedrite Group.
Weight.—5lbs.
Loc.—In Queensland, three miles N. of Mungindi Post Office,
        N. S. Wales.
Finder and Date.—Early in 1897.
Coll.—Mining and Geological Museum, Sydney.

MUNGINDI, No. 2.—G. W. Card, Rec. Geol. Surv. N.S. Wales,
        v., 3, 1897, p. 121.
Type.—Siderite, apparently part of the same meteorite as
        Mungindi, No. 1.
Weight.—62lbs.
Loc.—Found with Mungindi, No. 1.
Finder and Date.—Early in 1897.
Coll.—Mining and Geological Museum, Sydney.

NARRABURRA.—H. C. Russell, Journ. Roy. Soc. N.S. Wales, xxii.,
        1890, p. 81.
Type.—Siderite.
Weight.—70lbs. 14ozs.

Loc.—Narraburra Creek,* twelve miles E. of Temora, N.S.
Wales, Lat. 34° 10' S., Long. 147° 43' E.
Finder and Date.—Mr. O'Brien, 1854.
Coll.—H. C. Russell, F.R.S., Govt. Astronomer, Sydney.

NOCOLECHE.—T. Cooksey, Rec. Austr. Mus., iii., 3, 1897, p. 51.
Type.—Siderite, belonging to the Octahedrite Group.
Weight.—44·18lbs. (20,040 grams.)
Loc.—Five miles S.W. of Nocoleche Station, near Wanaaring,
N.S. Wales.
Finder and Date.—Existence known in 1895.
Coll.—Australian Museum, Sydney ; H. A. Ward, Rochester,
U.S.A.

ROEBOURNE.—
Type.—Siderite.
Weight.—191½lbs.
Loc.—Two hundred miles S.E. of Roebourne, N.W. West
Australia.
Finder and Date.—H. Reginald Hester, in 1892.
Coll.—Perth Museum, Perth, West Australia ; H. A. Ward,
Rochester, U.S.A.

TEMORA.—G. W. Card, Rec. Geol. Surv. N.S. Wales, v., 2,
1897, p. 52 ; A. Brezina, Ann. K.K. Naturhist. Hofmus.
Wien., x., 3 - 4, 1895, p. 288.
Type.—Siderite, belonging to the Octahedrite Group.
Weight.—
Loc.—Between Cootamundra and Temora, N. S. Wales.
Finder and Date.—Found by some miners about the year
1890.
Coll.—Fragments:—Mining and Geological Museum, Sydney ;
Hofmuseum, Vienna ; H. A. Ward, Rochester, U.S.A.

THUNDA.—A. Liversidge, Journ. Roy. Soc. N.S. Wales, xx., 1886,
p. 73 ; *ibid.*, xxii., 1888, p. 341 ; A. Liversidge, Proc.
Austr. Assoc. Adv. Sci., ii., for 1890 (1891), p. 387 ; A.
Brezina, Ann. K.K. Naturhist. Hofmus. Wien., x., 3 - 4,
1895, pp. 272, 283.
Type.—Siderite, belonging to the Octahedrite Group.
Weight.—137lbs.
Loc.—Thunda, Windorah, Diamantina District, Queensland·
Coll.—A. Liversidge, F.R.S., University, Sydney (main
mass).

---

* Mr. Russell tells me personally that Yeo Yeo Creek is its proper
locality.

YOUNDEGIN, No. 1.—L. Fletcher, Min. Mag., vii., 34, 1887,
        p. 121; A. Brezina, Ann. K.K. Naturhist. Hofmus.
        Wien., x., 3–4, 1895, p. 286.
Type.—Siderite, belonging to the Octahedrite Group.
Weight.—Four fragments weighing 25¾lbs., 24lbs., 17½lbs., and
        6lbs., and broken pieces 17lbs.
Loc.—Three-quarters of a mile N.W. from Penkarring Rock,
        about seventy miles E. of York, West Australia, Lat.
        31° 30′ S., Long. 117° 30′ E.
Finder and Date.—Alfred Eaton, Jan. 5th, 1884.*
Coll.—Two fragments in British Museum (Nat. Hist.).

YOUNDEGIN, No. 2.—J. R. Gregory, *Nature*, 1892, xlvii., 1204,
        p. 90; A. Brezina, Ann. K.K. Naturhist. Hofmus.
        Wien., x., 3–4, 1895, p. 286.
Type.—Siderite, belonging to the Octahedrite Group.
Weight.—382½lbs.
Loc.—Youndegin, West Australia.
Finder and Date.—Louis Knoop, in 1891.
Coll.—J. R. Gregory, London.

YOUNDEGIN, No. 3.—*Nature*, 1893, xlvii., 1220, p. 469; A.
        Brezina, Ann. K.K. Naturhist. Hofmus. Wien., x.,
        3–4, 1895, p. 286.
Type.—Siderite, belonging to the Octahedrite Group.
Weight.—2044lbs.
Loc.—Youndegin, West Australia.
Finder and Date.—Louis Knoop, in 1892.
Coll.—J. R. Gregory, London.

A CONTRIBUTION TO A BIBLIOGRAPHY OF AUSTRALIAN METEORITES.

ANONYMOUS.—The Meteor of the 14th April [Haddon Meteorite].—
        The Illustrated Australian News, 17th May, 1875, pp.
        68 and 74.
    „   [Youndegin Meteorite, No. 3].—*Nature*, 1893, xlvii.,
        1220, p. 469.

BERTHELOT (M.)—Nouvelles contributions à l'histoire des Car-
        bones, du Graphite et des Météorites.—*Ann. Chimie*,
        1873, xxx., p. 424.
    „   Nouvelles Contributions à l'histoire du Carbone.—*Compt.
        Rend.*, 1871, lxxiii., p. 494.

* Catalogue of Exhibits in the Western Australian Court of the
Colonial and Indian Exhibition, London, 1886, gives 1883 as the date
of discovery.

BREZINA (A.)—Die Meteoritensammlung des K. K. naturhistorischen Hofmuseums am 1 Mai, 1895.—*Ann. K. K. Natur-hist. Hofmus. Wien*, 1895, x., 3 – 4, pp. 235, 252, 267, 268, 272, 273, 285, 294, 301, 305, 306, 307, 340, 341, 344, 355.

BUCHNER (O.)—Die Meteoriten, etc., pp. 202 (8vo. Leipsig, 1863). [Cranbourne Meteorite, p. 198].

CARD (G. W.)—On the Occurrence and Classification of some New South Wales Meteorites.—*Rec. Geol. Surv. N.S. Wales*, 1897, v., 2, p. 49.

„ Mineralogical and Petrological Notes, No. 6.—*Rec. Geol. Surv. N.S. Wales*, 1897, v., 2, p. 121.

COOKSEY (T.)—The Nocoleche Meteorite, with Catalogue and Bibliography of Australian Meteorites.—*Rec. Austr. Mus.*, 1897, iii., 3, p. 51.

FLETCHER (L.)—On a Specimen of Meteoric Iron found at Youndegin, West Australia, in 1884.—*Min. Mag.*, 1887, vii., No 34, p. 121.

„ An Introduction to the Study of Meteorites.—*Brit. Mus. (Nat. Hist.) Guides*, 1886, pp. 45, 67.

FLIGHT (W.)—A Chapter in the History of Meteorites.--*Geol. Mag.*, 1875, (2), ii., pp. 264, 552.

„ Supplement to a Chapter in the History of Meteorites.—*Geol. Mag.*, 1882, (2), ix., pp. 107, 448; *Ibid.*, 1883, (2), x., p. 59.

„ Report of the Examination of the Meteorites of Cranbourne in Australia, of Rowton in Shropshire, and of Middlesborough in Yorkshire.—*Phil. Trans.* for 1882 (1883), clxxiii., p. 885.

FOORD (G.)—Nickeliferous and Meteoric Iron.—*Brough Smyth's GoldFields and Mineral Districts of Victoria*, 1869, p. 424.

GIBBONS (S.)—Note on the Cranbourne Meteorite.—*Trans. Roy. Soc. Vict.*, 1874, x., p. 130.

GREGORY (J. R.)—A Large Meteorite from West Australia.—*Nature*, 1892, xlvii., 1204, p. 90.

HAIDINGER (W. VON)—Die Dandenong Meteoreisenmasse in Melbourne.--*Sitz. K. Akad. Wiss. Wien*, 1861, xliv., pp. 378, 465; *Ibid.*, 1862, xlv., p. 63.

HAUSKOFER (K.)—Meteorit von Cranbourne, Australien.—*Journ. Prakt. Chem.*, 1869, cvii., p. 330.

„ Meteorite found near Cranbourne, Melbourne, Australia.—*Chem. News*, 1870, xxi., p. 12.

LE GOULD (L.)—Discovery of an Aerolite, and Visit to a Petrified Forest in Northern Queensland.—*Geol. Mag.*, 1864, i., p. 142.

LIVERSIDGE (A.)—The Deniliquin or Baratta Meteorite.—*Trans. Roy. Soc. N.S. Wales* for 1872 (1873), p. 97 ; *Journ. Roy. Soc. N.S. Wales* for 1882, xvi., p. 31.

,,   On the Bingara Meteorite.—*Journ. Roy. Soc. N.S. Wales* for 1882, xvi., p. 35.

,,   Metallic Meteorite, Queensland.—*Journ. Roy. Soc. N.S. Wales* for 1886 (1887), xx., p. 73.

,,   [The Thunda Meteorite].—*Journ. Roy. Soc. N.S. Wales* for 1888 (1889), xxii., 2, p. 341.

,,   Australian Meteorites.—*Proc. Austr. Assoc. Adv. Sci.* for 1890 (1891), ii., p. 387.

,,   The Minerals of New South Wales, etc.—(8vo. London, 1888), pp. 207, 218, 221.

MINGAYE (J. C. H.)—Notes and Analyses of a Metallic Meteorite from Moonbi, near Tamworth, N.S. Wales. *– Journ. Roy. Soc. N. S. Wales* for 1893, xxvii., p. 82.

RUSSELL (H. C.)—[Meteorite near Hay (Eli Elwah Meteorite)].— *Journ. Roy. Soc. N.S. Wales* for 1888 (1889), xxii., 2, p. 341.

,,   [The Baratta and Gilgoin Meteorites].—*Journ. Roy. Soc. N.S. Wales* for 1889, xxiii., 1, p. 46.

,,   [The Narraburra Meteor].—*Journ. Roy. Soc. N.S. Wales* for 1890, xxiv., p. 81.

,,   On Meteorite No. 2, from Gilgoin Station.—*Journ. Roy. Soc. N.S. Wales* for 1893, xxvii., p. 361.

.,   On a Meteorite from Gilgoin Station.—*Nature*, 1894, xlix., 1266, p. 325.

SELWYN (A. R. C.) AND ULRICH (G. H. F.)—Notes on the Physical Geography, Geology, and Mineralogy of Victoria.—(8vo. Melbourne, 1866). [Cranbourne Meteorite, p. 517].

SMITH (J. LAWRENCE).—On the Composition of the new Meteoric Mineral Daubréelite, and its frequent, if not universal occurrence in Meteoric Iron.—*Silliman's Amer. Journ. Sci.*, 1878, (3) xvi., p. 270; *Min. Mag.*, 1879, ii., 9, p. 152.

ULRICH (G. H. F.)—*Vide* Selwyn (A. R. C.) and Ulrich (G. H. F.)

WARD (H. A.)—Supplementary Catalogue of Meteorites for Sale, April 1, 1897.

WILKINSON (C. S.)—[The Cowra Meteorite].—*Journ. Roy. Soc. N.S. Wales* for 1888 (1889), xxii., 2, p. 341.

I am indebted to Prof. A. Liversidge, F.R.S., Messrs. H. C. Russell, F.R.S., R. T. Baker, G. M. Card, B. H. Woodward, and the Curator of this Museum, for assistance in compiling this Catalogue and Bibliography.

ANKERITE from SANDHURST, VICTORIA.

By T. Cooksey, Ph.D., B.Sc., Mineralogist.

Among the specimens of minerals in the Museum Collection are two, which were obtained from the New Chum line of reef, Sandhurst, Victoria, and which had been placed among those of the Calcite group. That these were correctly named seemed doubtful, as the powdered mineral effervesced very feebly with dilute hydrochloric acid in the cold. A qualitative test showed that both iron and magnesia were present in considerable quantity and a complete analysis furnished the following results :—

CaCO$_3$ ...	...	...	48·95
FeCO$_3$ ...	...	...	23·12
MgCO$_3$	...	...	25·01
Insoluble residue		...	3·54
			100·62

·3016 gram of material was taken for analysis, which on treating with hydrochloric acid, left ·0107 gram of insoluble matter consisting mainly of albite. By subtracting this insoluble portion from the total quantity taken, namely ·3016 gram, and calculating the results on the amount dissolved, that is ·2909 gram, the percentage composition of the three carbonates is found to be :—

CaCO$_3$ ...	...	...	50·76
FeCO$_3$ ...	...	:..	23·97
MgCO$_3$	...	...	25·93

Manganese was not present, neither the borax bead test nor the fusion with nitre and caustic potash giving the manganese reaction.

The specific gravity of the mineral is 2·994 (uncorr.) and its hardness about 3·5.

The crystals consist of very flat rhombohedrons with slightly curved faces occasionally striated, and form the lenticular crystals with sharp edges frequently seen in calcite and more especially in siderite. They are, however, externally slightly altered and the surfaces have a dull yellowish tinge deepening occasionally towards the edges. Internally they are colourless and translucent and show the rhomboidal cleavage perfectly.

In the one specimen the crystals of ankerite are associated with large and well formed crystals of quartz, some of which are left-handed, having both the rhombo- and trapezohedral surfaces. A few saddle-shaped crystals of siderite are deposited here and there on both minerals.*

The other specimen contains no siderite, the associated minerals being quartz crystals, and a few large and numerous small crystals of albite.

Normal ankerite has the formula $2\ CaCO_3.FeCO_3.MgCO_3$ assigned to it and requires

$2\ CaCO_3$		...	50·0
$FeCO_3$	...	...	29·0
$MgCO_3$	...	...	21·0
			100·0

The analysis of the above mineral is seen to differ from this in the relative proportions of the carbonates of iron and magnesia. Calculated however for the formula $5\ CaCO_3.2\ FeCO_3.3\ MgCO_3$ the percentages found agree exceedingly well with the theoretical

	Calculated.	Found.
$5\ CaCO_3$	50·81	50·76
$2\ FeCO_3$	23·57	23·97
$3\ MgCO_3$	25·61	25·93

Boricky† writes the formula for ankerite and similar minerals thus :—

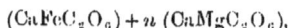

$$(CaFeC_2O_6) + n\ (CaMgC_2O_6),$$

and states that n may vary from $\frac{1}{2}$ to 10. When n is 1 the formula represents normal ankerite, but he assumes that those minerals in which n is 2 or less may be classed as ankerites, while he names those in which n is greater than 2 parankerites. The formula $5\ CaCO_3.2\ FeCO_3.3\ MgCO_3$ calculated for the present mineral may be written :—

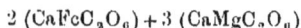

$$2\ (CaFeC_2O_6) + 3\ (CaMgC_2O_6)$$

---

* These crystals of siderite have already been figured and described by the late Mr. F. Ratte in " Notes from the Australian Museum," Proc. Linn. Soc. N. S. Wales, x., 4, 1885, p. 750.

† Boricky—Min. Mitth., xlvii., 1876.

and this again :

$$(CaFeC_2O_6) + \frac{2}{3}(CaMgC_2O_6)$$

in which it is seen that n is equal to $\frac{2}{3}$. This mineral therefore, according to Boricky's nomenclature, must be placed with the ankerites, but differs from the normal however in that n is $\frac{2}{3}$ instead of 1.

In Australia generally this mineral has been rarely observed. In New South Wales, Queensland, and Tasmania it has not been recorded. In Victoria, A. R. C. Selwyn and G. H. F. Ulrich[*] state that a mineral similar in composition was met with in veins and patches in decomposed basalt at Philip Island ; and in South Australia it has been found at Gill's Bluff, near Lyndhurst, and at the Wallaroo Mine. Brown spar and ferroanleite have been observed in several places, but they vary very considerably in composition from anker ite itself.

On a PRECAUDAL VERTEBRA of *ICHTHYOSAURUS AUSTRALIS*, McCOY.

By R. ETHERIDGE, JUNR., Curator.

THE subject of this paper is the imperfect vertebra of a large Ichthyopterigian, referable, I believe, to *Ichthyosaurus australis*, McCoy.* The original was brought under my notice by the Rev. M. Kirkpatrick, of Bega, N. S. Wales, who obtained it from Marathon, Central Queensland. With his permission a cast was taken for the Australian Museum Collection. As Sir F. McCoy's description was very brief, an extended notice of one of the middle trunk, or anterior pre-caudal vertebræ, may be acceptable to Australian investigators.

The specimen is the centrum of a large vertebra measuring five inches in its vertical and transverse diameters, and rivals in size those of the gigantic *I. campylodon*, Carter, from the European Chalk, the vertebra figured† by the late Sir Richard Owen measuring only four inches high. Our example is devoid of the neural spine, neurapophyses, and pleurapophyses, but having the articular surfaces of the first and last well displayed. The positions of the diapophysial and pleurapophysial articular surfaces leads to the belief that the vertebra is one of the middle trunk series. It is subcircular in outline, slightly narrowed and contracted neurally. Measured across the articular surfaces from the neural to the hæmal margins the diameter is exactly five inches, and in a transverse direction, from diapophysis to diapophysis it is an eighth of an inch short of a similar measurement. Longitudinally measured between the dia- and pleuraphysial tubercles the centrum is exactly two inches, but on the hæmal surface it is a quarter of an inch more.

The concave terminal articular surface visible is deep, terminating in a central fossa, the extent of the concavity being well exemplified by the matrix cast of the anterior cavity of the succeeding vertebræ at the posterior end of this specimen. This mass of matrix represents the " elastic capsule " that intervened between the vertebræ, and retains on its surface portions of the osseous tissue of the succeeding centrum. The periphery or immediate articular rim at each end is narrow, the surface thence sloping rapidly inwards, but the edges of the rims project slightly

---

* Trans. Roy. Soc. Vict., viii., 1868, p. 41.

† Owen—Mon. Foss. Reptilia Cret. Formation, p. 79, pl. xxii.

outwards, thus rendering the longitudinal or lateral surfaces of the centrum somewhat concave. The depth of the concavities is an inch, or perhaps a little more, and a longitudinal section of the centrum would be, in consequence, of a strongly hour-glass shaped outline. The floor of the myelonal canal is three-quarters of an inch wide, the joint faces of the neurapophysial surfaces rather triangular on very strongly raised fore and aft synchondrosial articular elevations, the space between these and the diapophysial tubercles is roughly three inches, the latter having descended in close contiguity to the parapophysial tubercles. It is clear, therefore, that this vertebra from the wide disassociation of the neura- and diapophyses occupied a position in the column certainly more than one-third of the trunk from the head, and, according to Owen's measurements, was near about the forty to forty-fifth vertebra, for in this region in *Ichthyosaurus*, the dia- and para-pophyses form a pair of separate tubercles on each side near the anterior ond of the centrum.

The diapophyses are set further in from the anterior articular edge than the parapophyses ; these are close to the latter, but are not connected with it by a "neck." Both are represented by large and strong rounded tubercles, separated from one another by an interval of two-eighths of an inch, this interstitial surface being deep and groove-like. The hæmal surface is quite plain.

The posterior concave articular surface is infilled with matrix, affording a complete cast of the next succeeding anterior cup, and even retaining a portion of the osseous tissue of the latter adhering to it. This tissue throughout the centrum is well preserved and dense.

The specimen is certainly of the Campylodont group of *Ichthyo-sauri*, and occupies an intermediate position in outline between an "early posterior dorsal" and a "late posterior dorsal" vertebra of *I. trigonus*, Owen.*

The largest *Ichthyosauri* attained a length of from thirty to forty feet, and the present meridianal species was in no way inferior to its gigantic fellows of the European Secondary seas. If we apply a similar method of arriving at the comparative size of an *Ichthyosaurus* as that adopted by Prof. Owen, that the jaw was "thirteen times the length of the vertical diameter of an abdominal or anterior caudal centrum," we see in the present fossil the representative of an animal possessing a jaw a little over five feet in length—thus $13 \times 5'' = 65'' = 5' 5''$ long. Prof. McCoy computed† the remains of his type specimen to represent

---

* Lydekker—Brit. Mus. Cat. Foss. Reptilia and Amphibia, 1889, pt. 2, p. 26, figs. 13 and 14.

† Trans. Roy. Soc. Vict , ix., 2, 1869, p. 77.

an animal twenty-five feet long. Similar vertebræ to that now
described he states possessed a diameter of four inches, and elsewhere he remarks* that the longitudinal measurement reached
one and a half inches. The elastic capsule was also preserved in
some of his specimens.

Mr. R. Lydekker, in the previously mentioned "Catalogue,"
gives a list of species that "cannot be classified."† Amongst
these are *I. australis*, McCoy, and *I. marathouensis*, mihi. I am
afraid he has overlooked Sir Frederick's principal paper on his
I. australis, wherein, although the description is meagre, the latter
specially compares the teeth of his fossil to those of *I. campylodon*,
and says they "have a rough bony square base like those of *I.
campylodon* (Carter)." As regards *I. marathouensis*, mihi, less
can perhaps be definitely said, but the whole of its structure, so
far as we know it, is also after the type of *I. campylodon*. In my
paper on this fossil, I called attention to the necessity of affording
another name to *I. australis*, Hector, a New Zealand species
distinct from McCoy's. This has now been done by Mr. Lydekker
terming it *I. hectori*,‡ but unfortunately the species is of no value,
from the absence of either description or figure, all that Sir James
Hector says about it being "this genus is only represented in the
collection by a single vertebral centrum."

Ichthyosaurus indicus, Lydk.,§ seems to be an allied species to
I. australis, and also vied with *I. campylodon* in size. It is from
the Ootatoor Group, the homotaxial equivalent of the Chalk Marl
and Upper Greensand of England.

McCoy's original specimens were from Walker's Table Mountain
on the Flinders River. The present vertebra is, as before said,
from Marathon on the same stream. Both are localities in the
Rolling Downs Formation, or Lower Cretaceous.

---

* Ann. Mag. Nat. Hist., (3), xix., 1867, p. 355.
† *Loc. cit.*, p. 113.
‡ *Loc. cit.*, p. 113.
§ Pal. Indica (4), i., 3, 1879, p. 27.

A NEW BLIND SNAKE FROM THE DUKE OF YORK ISLAND.

By EDGAR R. WAITE, F.L.S.,
Zoologist, Australian Museum.

THE species herein described, is represented by two specimens, collected, together with other material, in the Duke of York Island, by the Rev. George Brown, D.D., and forwarded to the Museum many years ago. For this snake I propose the name of—

TYPHLOPS SUBOCULARIS, sp. nov.

Habit stout, of even thickness. Head flat, much depressed. Snout prominent, with a rather acute, but not sharp, edge. Rostral above, very narrow, about one-sixth the width of the head, reaching but little more than half-way to the level of the eyes ; the portion visible from beneath somewhat longer than broad ; nasal incompletely divided, the fissure extending from the anterior portion of the second labial. Nostrils inferior, preocular, but little narrower than the ocular, separated from the labials by one scale, ocular separated by two scales. Four upper labials, the anterior three very narrow. The mandible is ∧-shaped, the symphysis very acute, and the scales adjoining the labials oblique. Diameter of the middle of the body thirty times in the total length. Tail longer than broad, terminating in a large thorn-like spine. Thirty-six (specimen A.) or thirty-four (specimen B.) scales round the middle of body.

Colours.—In spirits, dark brown above, yellow beneath, the two areas strongly marked, line of union interrupted.

Dimensions.

	A. (Type).	B.
Total length	360·0 mm.	280·0 mm.
Length of head	7·5 ,,	6·0 ,,
Width of head	9·0 ,,	7·5 ,,
Width of body	12·1 ,,	9·2 ,,
Length of tail	17·3 ,,	7·8 ,,
Width of tail	11·5 ,,	6·5 ,,

Habitat.—Duke of York Island. Two specimens.

Type.—In the Australian Museum, Sydney.

This species will come into Division III. of Boulenger's synopsis,[*] "a præocular and one or two suboculars," but will fit into neither of the subordinate groups, the character of the snout not being sufficiently marked. Taking the number of scales round the body, namely, thirty-four to thirty-six, the only described species with which it thus agrees is *T. acutus*,[†] at once distinguishable (apart from its distribution) by the peculiar snout and the remarkable size of the rostral.

A species, *T. depressus*,[‡] has previously been made known from the Duke of York Island ; in this the præocular is in contact with the labials, and a comparison of Peters' figures (a reference to which Mr. Boulenger has, in his Catalogue,[§] omitted) with those below given, shows how widely the two species differ.

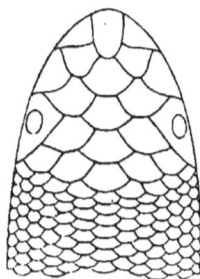   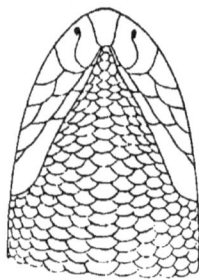

1.

From above.

2.

From below.

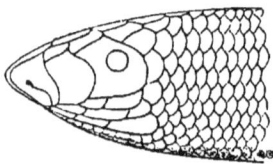

3.

In profile.

Head of *Typhlops subocularis* (enlarged).

---

* Boulenger— Brit. Mus. Cat. Snakes, i., 1893, p. 14.
† Dum. et Bibr., Erpét. gén., vi., 1844, p. 333.
‡ Peters—Monatsb. K. Preus. Akad. Wiss. Berlin, 1880, p. 220, fig. 3 (p. 309).
§ Boulenger—*loc. cit.*, p. 33.

NEW OR LITTLE-KNOWN LOWER PALÆOZOIC GASTEROPODA IN THE COLLECTION OF THE AUSTRALIAN MUSEUM.

By R. ETHERIDGE, JUNR., Curator.

(Plates xv., and xvi.)

THE following Lower Palæozoic fossils are either new to Australian Palæontology, or have been imperfectly described.

Genus Goniostropha, *Œhlert*, 1888.

(Bull. Soc. Études Sci. Angers for 1887 (1888).)

GONIOSTROPHA PRITCHARDI, *sp. nov.*

Pl. xv., Figs. 1 – 4.

Sp. char.—Shell small, elongately spiral, of seven or eight slowly increasing angular whorls, each bearing two sharp median keels, enclosing between them a smooth, slightly concave band, and the remaining surface of each whorl occupied by a series of fine spiral lyræ that are sometimes finer above than below the band ; sutures deep ; mouth with the outer lip apparently rounded, and the inner lip reflected. Length (of largest specimen) one and a quarter inches.

Obs.—None of the mouths in the specimens before me are perfect, but the outer lip seems to have been rounded in outline, and the inner lip is certainly reflected. The whorls are only rendered angular by the projecting principal keels, enclosing between them the band, and they become less median in position as the apex is approached. This form belongs to a group of rather common *Murchisonia*-like shells for which Dr. Daniel Œhlert has proposed the name *Goniostropha*, distinguished by more or less angular whorls, the band occupying the angle or greatest periphery of each whorl, accompanied by supplementary finer revolving lyræ. In this respect it differs from an allied genus, *Hypergonia*, Donald.[*]

As this is an undescribed species from the Lilydale Limestone of Victoria, I have much pleasure in associating with it the name

---

[*] Quart. Journ. Geol. Soc., 1889, xlv., p. 623.

A

of Mr. G. B. Pritchard, who has done much to advance our knowledge of Victorian Palæontology.

Loc. and Horizon.—Cave Hill Quarries, Lilydale, Upper Yarra District, Victoria. Lilydale Limestone, Upper Silurian ; presented by Messrs. J. Hinder and E. Smith.

Genus Gyrodoma, *gen. nov.**

GYRODOMA ETHERIDGEI, *Creswell, sp.*

Pl. xvi., Fig. 1.

Eunema etheridgei, Creswell, Proc. R. Soc. Vict., 1893, v. (n.s.), p. 42, t. 8, f. 2, (2 figs.).

Obs.—Mr. Creswell's figure represents an imperfect shell, and those now before me are also in the same condition, but sufficient characters are deducible to show that it is not referable to *Eunema,* which is an imperforate genus, with angular whorls. In *G. etheridgei,* on the other hand the whorls are boldly rounded, certainly seven in number and possibly more ; in Pl. xvi., Fig. 1 seven are visible, whilst Creswell assumed five to be the number. In the latter's left hand figure, as above quoted, there seems to me to be an umbilicus, although no mention is made of this in the description. In Pl. xvi., Fig. 1, a distinct and rather flat or depressed band is visible, bounded by two lyræ that are certainly more prominent than the remainder encircling the whorls. Mr. Creswell's figures both distinctly portray two bands side by side, but the example now figured has but one. If a double band does exist on some specimens, it indicates a departure towards the Cretaceous genus *Disopeta,* Gardner. The whorls decrease in size rapidly, the inner lip is almost straight, and with the anterior termination of the outer lip describes nearly a right angle.

The presence of the regular spiral lyræ distinguishes the proposed new genus from the allied *Goniostropha,* Œhlert, *Cœlocaulus,* Œhlert, *Hormotoma,* Salter, *Caliendrum,* Brown, *Cerithioides,* Haughton, *Stegocœlia,* Donald, *Hypergonia,* Donald, and *Glyphodeta,* Donald, aided in some instances by the rounded whorls of *Gyrodoma,* and the position of the band. The nonderolement of the whorls distinguishes the latter from *Loxoplocus,* Fischer, and the absence of all tuberculation from *Murchisonia,* the genus proper in its restricted sense. If an umbilicus exists it cannot be a *Pithodea,* De Kon. I therefore conclude it is a new section of *Murchisonia* and call it *Gyrodoma.*

Loc. and Horizon.—Cave Hill Quarries, Lilydale, Upper Yarra District, Victoria. Lilydale Limestone, Upper Silurian ; presented by Mr. J. Hinder.

* *Gyro,* to turn round, and *domus,* a house.

Genus Mourlonia, *De Koninck*, 1883.

(Faune Calc. Carb. Belgique, 1880, pt. 4, p. 75.)

MOURLONIA DUNI, *sp. nov.*

Pl. xv., Fig. 5 ; Pl. xvi., Fig. 2.

Sp. char.—Shell conical, or somewhat trochiform, the sides of the cone fairly continuous; spire rather depressed, but acute at the apex ; whorls six, gently rounded ; base convex. Body whorl large, more than twice the height of the penultimate whorl, obtusely angular at the centre ; sutures faintly impressed ; band sutural on all but the body whorl, bounded above by a faint keel, on the body whorl occupying the obtuse median angle, the bounding keels very sharp and distinct, with a faintly impressed groove below the lower, and apparently without special sculpture ; sinus unknown ; umbilicus open, although not widely so ; mouth oval, with the inner lip reflected somewhat over the umbilicus, but without concealing it.

Obs.—The late Mr. Felix Ratte figured* three univalves from our Lower Palæozoic rocks without assigning specific names to them, nor even generic in the case of two. Whether or no the present shell be one of these I am in doubt, but in some points his fig. 6 is like it, and again in other respects widely divergent ; for instance in the figure quoted there is too great a convexity of the whorls, too elevated a spire, and too prominent a band. At the same time there is the possibility that the two may be identical, allowing for defective drawing.

Mourlonia duni is an exceedingly characteristic species of the Wellington Caves Limestone, and is at present unknown to me from any other horizon. It is named in honour of my former Assistant, Mr. W. S. Dun, to whom I am indebted for much cordial help.

Loc and Horizon—Wellington Caves, N. S. Wales. Siluro-Devonian.

Genus Helicotoma, *Salter*, 1859.

(Canadian Organic Remains, 1859, Dec. I., p. 10.)

HELICOTOMA JOHNSTONI, *sp. nov.*

Pl. xv., Figs. 6 – 8 ; Pl. xvi., Figs. 3 and 4.

Straparollus (Maclurea) tasmanicus, Johnston, Geol. Tas., 1888, t. 5, f. 7 (excl. f. 1 and 1a).

Sp. char.—Shell discoid, of about four whorls, each nearly twice the breadth of the preceding ; spire short, wholly depressed

---

* Proc. Linn. Soc. N. S. Wales, x., 1, 1885, t. 9. f. 6.

below the level of the body whorl, which is traversed on its outer
angle by a keel, without nodes, crenulations or echinations, but
variable in its degree of prominence and acuteness, on the inner
side of the keel the surface of the whorl slopes rapidly away to
the suture, with immediately above it a second feeble obtuse keel;
on the under side the surface of the body whorl is either gently
rounded or flattened, but the inner whorls rounded only ; umbilicus
telescopic, but most of the whorls visible ; back in no way concave
beneath the keel of the body whorl, but rounded and broadening
towards the mouth, the surface loosing much of its convexity.
Mouth generally rounded, but slightly insinuated at the keel, and
more so along the sutural margin, the upper margin, in other words,
retreating towards the shallow notch or insinuation referred to,
and the lower edge advancing and insinuated. Sculpture of the
upper surface consists of fine obliquely retreating lyrulæ on the
inner half of each whorl, and similar advancing lyrulæ on the
outer half, giving a faintly V-shaped figure, but on the wide
back these lyrulæ become more directly transverse ; on the under
surface the lyrulæ describe a sigmoidal curve, becoming much
coarser and sub-laminar towards the mouth, the sharpest portion
of the curve being immediately above the suture at the obtuse
feeble keel.

Obs.—Under the name of *Straparollus tasmanicus,* I feel convinced Mr. R. M. Johnston has included two perfectly distinct
shells. His figs. 1 and 1a. represent a *Raphistoma* that may be
known as *Raphistoma tasmanicum,* Johnston, *sp.,* whilst fig. 7,
the subject of the present remarks, appears to me to be a *Helicotoma,* and I therefore propose for it the name of *H. johnstoni.*
The specimen now figured was received in a collection of fossils
forwarded from the Tasmanian Museum. Pl. xv., Fig. 6, represents the upper side, corresponding to Johnston's fig. 7, whilst
our Pl. xv., Fig. 7, is that of the under or umbilical side of the *same*
specimen. I am puzzled how to distinguish this from another
shell that Mr. Johnston has figured as *Lituites,* sp. indet.,[*] except
that in the latter the transverse laminæ are too coarse for the lines
occurring on the under surface of *Helicotoma johnstoni.*

In the faintly V-shaped outline of the sculpture on the upper
side, the apex of the V is at the keel of the body whorl, producing
a slight notch on the outer lip, after the manner of *Helicotoma,*
without in anyway simulating a true sinus. This reflection of the
sculpture lines and the presence of the notch distinguish this
shell at once from *Polytropis,* De Koninck.

Loc. and Horizon.—Gordon River, West Tasmania. Gordon
River Limestone, Lower Silurian.

---

[*] Johnston—Geol. Tas., 1888, t. 5, f. 8 and 10.

Genus Trochonema, *Salter,* 1859.

(Canadian Organic Remains, 1859, Dec. 1, p. 24).

TROCHONEMA ETHERIDGEI, *Johnston.*

Pl. xvi., Figs. 5 and 6.

Trochonema etheridgei, Johnston, Geol. Tas., 1888, t. 5, f. 13 and 14.

Sp. char.—Shell turbinate, of five or six acutely keeled and angular whorls, the principal keel occupying the periphery of each whorl; on the anti-penultimate whorl there are three keels, the first small and thread-like bordering the upper suture, the surface thence to the second keel being tabulate or flat, from the latter to the principal keel slightly oblique and concave, and thence to the lower suture the surface is straight-walled; the penultimate whorl possesses four keels besides the peripheral, the three upper arranged as in the anti-penultimate, whilst between the third and fourth the surface is again straight-walled; the body whorl (somewhat hidden in matrix) probably possessed four also, the peripheral keel being strong and prominent. Sutures excavated. Mouth almost rhomboidal; outer lip strongly angled at the peripheral keel, rounded below; inner lip possibly straight. Umbilicus distinct. Sculpture consisting of oblique sub-imbricating growth lamellæ, faintly varicose along the peripheral keel, and becoming much stronger and rugose towards the mouth on the body whorl.

Obs.—This well marked shell was figured but not described by Mr. Johnston. In his plate explanation the author remarks that *T. etheridgei* is allied to *T. tricarinata,* Meek,* of the Corniferous Group of North America. *T. tricarinata,* Meek, should be known as *T. meekianum,* Miller.† The present shell is readily distinguished from *T. montgomerii,* mihi, by its much more turbinate form, and different arrangement of the spiral keels and sculpture.

Loc. and Horizon.—Gordon River, West Tasmania. Gordon River Limestone, Lower Silurian.

TROCHONEMA MONTGOMERII, *Eth., fil.*

Eunema montgomerii, Eth., fil., Ann. Rep. Secy. for Mines Tas. for 1895-6 (1896), p. xlvii., pl. f. 21 and 22.

Obs.—Since the publication of this species, further examples have been received from the Tasmanian Museum, one with an umbilicus exposed. This will necessitate its removal from *Eunema* to *Trochonema.* The following additional features may be noted :

---

* Ohio Geol. Report, Pal. I., 1873, p. 218, t. 19, f. 5 *a* and *b*.

† N. American Geol. and Pal., 1889, p. 428.

Each thread-like lyrula of the sculpture is separated from its neighbour by several times its own thickness, the obliquity of the lyrulæ on the upper part of each whorl being changed on the straight-walled portion to a perfectly vertical direction. The upper part of the inner lip, although not forming a callosity, is revolute, slightly projecting over the umbilicus. The aperture was long oval, angled on the outer lip by the principal keel of the body whorl.

Loc. and Horizon.—Gordon River, West Tasmania. Gordon River Limestone, Lower Silurian.

TROCHONEMA ? NODOSA, *sp. nov.*

Pl. xv., Figs. 9, 10.

Worthenia, sp. nov., Ratte, Proc. Linn. Soc. N.S. Wales, x., 1885, pt. 1, p. 80, t. 9, f. 1 and 2.

Sp. char.—Shell turbinate, but not depressed ; whorls more than four (four in part only preserved), the body whorl apparently not free, each whorl horizontal or nearly so on its upper portion around the suture, vertical or straight-walled in the lower ; all, except the body whorl, bear two keels, the latter three, the uppermost keel in each case demarcating the two portions of the whorls, and carring a number of blunt nodes, or tubercles, which on the body whorl become of a variciform nature, and more pronounced with the growth of the whorl ; the second keel is midway between that just mentioned and the suture, and with the third on the body whorl is nodose also. Mouth generally oval, vertically elongated ; outer lip quadrangular ; inner lip and minute sculpture not preserved ; umbilicus deep and apparently open.

Obs.—Had not Mr. Ratte figured this shell, and referred it to *Worthenia* (with which it has no connection), without a specific name, I should not have noticed it in consequence of its poor state of preservation. I am even doubtful of its proper generic resting-place from the same cause, but *Trochonema,* so far as I can judge, seems to be the most appropriate genus. At the same time it departs from the majority of species referred to the latter by the nodose nature of the encircling keels. There is one species of this genus, however, similarly ornamented—*T. yandellana,* Hall & Whitfield,[*] from the North American Corniferous Limestone, but otherwise distinct from *T. ? nodosa.* It may even be related to our old friend *Buccinum breve,* Sby., of the British Devonian rocks, and which Whidborne has of late referred[†] to the recent genus *Liotia,* Gray, without, however, in my opinion, sufficient reason.

---

[*] 24th Ann. Rep. N. York State Cab., 1872, p. 194 ; 27th *ibid.*, 1875, t. 13, f. 3 ; Nettelroth, Kentucky Fossil Shells, 1889, t. 20, f. 3.

[†] Mon. Dev. Fauna S. England, 1892, pt. 4, p. 271.

Loc. and Horizon.—Cave Flat, Murrumbidgee River, N. S. Wales. Cave Flat Limestone, Siluro-Devonian.

Genus Holopea, *Hall*, 1847.

(Pal. N. York, 1847, i., p. 169).

HOLOPEA WELLINGTONENSIS, *sp. nov.*

Pl. xv., Fig. 11 ; Pl. xvi., Fig. 7 9.

(*Unnamed shell*), Ratte, Proc. Linn. Soc. N.S. Wales, x., 1885, pt. 1, t. 9, f. 3 – 5.

Sp. char.—Shell ventricose, of six whorls rapidly decreasing in size above the last or body whorl ; apex acute ; whorls rounded, uniformly ventricose, or almost inflated, very much wider than high, slightly horizontally flattened around the suture ; body whorl very much expanded in proportion to the others. Mouth round ; outer lip sharp and fine ; inner lip straight and slightly thickened. Umbilicus open and deep. Sculpture consisting of a large number of regular, fine, equidistant, sharp revolving threads crossed by others exceedingly fine and oblique, giving rise to a beautiful and minute cancellation ; towards the outer lip are a few coarse sub-laminar ridges.

Obs.—This species was figured, although neither named nor described, by the late Mr. Ratte, but his figure shows a revoluted inner lip that is not present in any of our specimens.

A similar flattening of the whorls around the suture is seen in *Holopea obesa*, Winchell.*

H. wellingtonensis is not unlike some forms of *Callonema*, Hall, but possess spiral and growth threads, instead of the latter only, and also lacks the obtuse angularity on the anterior part of the body whorl that is almost always seen in species of Hall's genus. Both *Holopea* and *Callonema* are umbilicated.

Loc. and Horizon.—Wellington Caves, N.S. Wales. Siluro-Devonian.

---

* Geol. Wisconsin, 1873 – 79, iv., 1882, p. 348, t. 27, f. 4.

HALYSITES IN NEW SOUTH WALES.

By R. ETHERIDGE, Junr., Curator.

(Plate xvii.)

THE history of this genus, not only in N. S. Wales, but in Australia generally, is a very brief one. *Halysites* was first recorded by the late Prof. L G. de Koninck,[*] who recognised *H. escharoides*, Lamk., in the collection of N. S. Wales fossils sent him by the late Rev. W. B. Clarke for determination. In the description given, however, there are no characters that would readily differentiate between this species, and the typical *H. catenulatus*, Linn. De Koninck gives Wellington as the locality, but I have never seen a *Halysites* from the limestone of that district.

Although the presence of this old Palæontological landmark has been in a generic sense, recorded as occurring in N. S. Wales, the microscopic structure has not been investigated, so far as I know.

The largest fasciculo-reticulate corallum that has come under my notice is a specimen measuring six and a half inches by four and a half, forming, in every case, a lax and spreading mass rather than a high erect growth. The intersecting reticulations, or "fenestrules," are very variable in size and shape, but always polygonal, the smaller having an average size of three by three mm., the largest observed fifteen by five mm., fourteen by twelve, and so on, with intermediate gradations, the angles of junction of the vertical laminæ or plates being equally variable. The reticulations are usually longer in one direction than another, but not by any means invariably so. The walls are strong, but in consequence of the alteration that has taken place, the epitheca on the free sides of the laminæ is rarely discernible, but when so, is well developed. The number of corallites on any one side of a reticulation varies from two to twelve, but the average number is from four to six.

Examination in thin sections renders the great amount of alteration the corallum has undergone apparent, an unfortunate circumstance common to a large number of our Lower Palæozoic

---

[*] Foss. Pal. Nouv.-Galles du Sud, pt. 1, 1876, p. 16; Clarke, Sed. Formations N. S. Wales, 4th Edit., 1878, p. 129.

corals, still sufficient details can be made out to elucidate the finer characters of our *Halysites.*

The mineral condition is very remarkable. The corals are preserved in a dark blue limestone, the tissues where unaltered being composed of the usual dark grey or brown sclerenchyma, the general infilling of all the intertabular spaces or old visceral chambers, being crystalline or granular calcite, the former in places with cleavage. Every here and there, however, the walls of the corallites are converted into a radiating siliceous mineral, or blebs of the same look as if forced into the walls; there is every reason to believe that the latter is chalcedony, in the form of Beekite rosettes, a by no means uncommon mineral in our Lower Palæozoic Invertebrata. In some cases these blebs occupy spaces within the corallites, breaking up the uniformity of the tabulate structure in a very marked manner.

Notwithstanding this excessive alteration the external walls are quite discernible, and here and there the continuous epitheca on both the free sides of the laminæ is visible also. As described[*] by Nicholson, the epitheca does not take any part in the " form-ation of the partition which actually divides any tube from its neighbour on either side," but the partitions are formed solely by the coalescent *walls* of the two contiguous corallites." Further-more, the corallites are of two orders, as in the well known *Halysites catenulatus,* Linn., thus at once distinguishing it from *H. escharoides,* Lamk. The larger, or normal corallites are oval, from three-quarters to one mm. in longest diameter, and the latter in the direction of the chain. In a macroscopic examination these may be at once distinguished by an outward bulging of the epithecated walls. The smaller corallites, or those of the second order, are ranged alternately with the larger, and are either round or quadrate, and each is enclosed by a thick wall of its own, distinct from the common or enclosing wall of the laminæ. The position of these "interstitial tubes," as they are termed by Nicholson, is equally discernible externally by a biconcavity of the wall opposite to each secondary corallite. The angles of junction of any two laminæ that assist in forming a reticulation, or fenestrule, are always occupied by an interstitial corallite, which, in well weathered specimens, is visible with an ordinary pocket lens. Septa are absolutely wanting.

The two sets of corallites become even more apparent in a vertical section. The normal tubes are *closely* tabulate, the tabulæ horizontal, very regular, and equidistant, five in the space of one mm., enclosing between them more or less transversely elongated or quadrangular intertabular spaces. The interstitial tubes, on the other hand, are very narrow and pipe-like, *sparsely*

---

[*] Nicholson—Tab. Corals Pal. Period, 1879. p. 227.

tabulate, the tabulæ far apart (they are nearly half a mm. apart), although complete, the intertabular spaces vertically elongate, and both the walls and tabulæ greatly thickened, (Pl. xvii., fig. 7) as compared with the similar parts of the normal corallites. There is not the slightest trace whatever of the small projections in the interstitial tubes, "apparently of a septal nature," discovered and described by Nicholson.*

The presence of the interstitial corallites, and the absence of septa clearly places the present coral within the group of *H. catenulatus*, the "Chain-coral" of the Wenlock, but as compared with the latter there is this remarkable difference. In *H. catenulatus* the tabulæ of the normal corallites are "comparatively remotely disposed," whilst in the interstitial tubes they are "much more numerous and more closely set."† In our coral, which I propose to call *Halysites australis*, the exact opposite is the case.

It is customary with Monographists to include in *H. catenulatus* a large number of other forms, in former days regarded as separate species, purely from external characters, but I think before this indiscriminate lumping is done, the whole of such forms should be submitted to microscopic examination, when possibly differences of an equally important nature to that shown above, may be found to exist.

Halysites australis differs from *H. agglomerata*, Hall,‡ by the same characters that it does from *H. catenulatus*, Linn. In the form of the reticulations and mode of growth it is more akin to *H. labyrinthica*, Goldf.,§ and *H. catenulatus*, var *Harti*, Eth.,‖ but in both cases the fenestrules are larger than in our species.

The specimens were collected by the Rev. J. Milne Curran, at Molong, N. S. Wales, and by him presented to the Trustees.

---

* *Loc. cit.*, p. 229, t. 11, f. 1.
† *Ibid.*, p. 228.
‡ Hall—Pal. N. York, ii., 1852, t. 35 bis.
§ Petrefacta Germaniæ, t. 25, f. 5a.
‖ Quart. Journ. Geol. Soc., xxxiv., 1878, p. 583, t. 28, f. 2.

DESCRIPTION OF THE LARVA OF *PSEUDOTERPNA PERCOMPTARIA*, Gn.

By W. J. Rainbow, Entomologist.

(Plate xviii.)

Family Geometridæ.

Sub-Family Boarmiinæ.

Genus Pseudoterpna, *Meyr.*

Pseudoterpna percomptaria, *Gn.*

(Pl. xviii., Figs. 1, 1*a*, 1*b*, 1*c*, 1*d*.)

Dorsal surface sage green with small black spots ; sides concolorous, with narrow longitudinal median stripes of pale yellow, the latter bordered with green ; lateral surfaces sparingly dotted with minute black spots ; in addition to these, the fourth, fifth, sixth, seventh and eighth segments have each a small red spot, seated just below the spiracular oriface ; spiracles white, oval, ringed with black ; ventral surface concolorous, with median stripe of pale yellow.

The *head* consists of an elongated, hard, chitinous process, wedge-shaped, deeply grooved down the centre both above and underneath ; back of head sage-green at base, apex suffused with pink and tipped with black, the surface finely granulated, and sparingly dotted with black ; sides granulated, sage-green, with a median longitudinal line of dark brown commencing at apex, and terminating rather lower down than half-way ; in front sage-green at base, suffused with pink, and thickly furnished with minute brown granules ; mouth parts of a dingy pinkish colour.

Legs short, closely grouped together, pale yellowish ; pro-legs small, grouped together, and attached to the two final segments.

Anal segment terminating with an elongated, chitinous, bifurcated wedge-shaped process, the surface of which is granulated ; sage-green at base, black at tips.

The specimen described was obtained by Mr. Sydney L. Evans, at Guyra, near Inverell, and was forwarded by him to the Australian Museum, where it was received on the 12th of April, 1897. Two days afterwards it entered the pupal stage. In the interval, however, that elapsed between the date it was received and the time of its pupating, the notes and sketches necessary for a description of the creature were made. The pupa was attached to a stick by its tail, and had a silken girdle across the middle, in which respect it might easily have been mistaken for a *Papilio.* This feature, it is well known, is not uncommon among the moths of the family Geometridæ.

In answer to a query as to its food-plant, Mr. Evans writes us as follows :—" I am sorry to say that I am unable to tell you what the food plant of the caterpillar is. I found it holding on to a blade of grass near the Guyra Lagoon, and at first sight took it for a folded leaf, but on closer inspection found it was alive, but could not decide 'which end was which,' as there was apparently no difference." Mr. G Lyell, Junr., informs me that he has observed the larva feeding on the Peppermint Gum *(Eucalyptus piperita,* Sm.)

The moth bred out on 3rd January, 1898.

DESCRIPTION of a NEW ARANEIAD.

By W. J. Rainbow, Entomologist.

(Plate xviii.)

Family Argiopidæ.

Genus Poltys, *C. Koch.*

Poltys multituberculatus, *sp. nov.*

(Pl. xviii., Figs. 2, 2*a,* 2*b.*)

♀ Cephalothorax 6·2 mm. long, 4·8 mm. broad ; abdomen, 11·6 mm. long, 8·8 mm. broad.

Cephalothorax longer than broad, arched, dark brown, almost black, glossy. *Caput* arched, terminating in front with a tubercular ocular eminence, the latter 0·5 mm. high, and clothed in front with long grey hairs ; commencing at base of ocular eminence and extending thence to the junction of the cephalic and thoracic segments there is, in the median line, a very conspicuous scopula, the hairs of which are long and grey. *Clypeus* broad, strongly arched, dark brown, almost black, glossy, median depression deep, radial grooves indistinct. *Marginal band* broad, and of a pale fleshy tint.

Eyes black ; of these six are seated on the tubercular ocular eminence, and arranged in two rows, the lower consisting of four eyes, and these are in a curved line directed downwards and forwards ; of this series the median pair are sensibly the largest of the group ; each eye of the anterior row is separated from each other by a space equal to fully twice their individual diameter ;

the two comprising the second row are separated from their anterior neighbours by a space equal to rather less than twice their diameter, and from each other by about three diameters; the remaining two lateral eyes are located in the angles of the cephalic segment at a distance from the tubercular eminence of about 0·5 mm.

Legs long, robust, armed with strong spines, yellow-brown with dark annulations, clothed on the outer margins with hoary grey hairs, on the inner margins with ferruginous grey; coxæ densely clothed underneath with grey hairs.   Relative lengths 1 = 2, 4, 3.

Palpi long, similar in colour, armature, and clothing to legs.

Falces long, robust, arched, glossy, sparingly clothed with black hairs, apices divergent, yellow at base to about one-third their length, where it terminates suddenly, and is thence dark brown to tips; the upper margin of the furrow of each falx is armed with a row of four teeth, of which the two nearest the base are the longest and strongest; the lower margin is armed with two.

Maxillæ broad, arched, moderately long, divergent; laterally they are yellowish-brown, and clothed with rather long, hoary greyish hairs; inner surfaces glossy, pale flesh-coloured and naked, but the edges are furnished with dense hoary scopulæ.

Labium broader than long, arched, obtuse at apex, yellow-brown from base to about one-half its length, thence pale flesh-coloured.

Sternum shield-shaped, moderately arched, densely clothed with short greyish hairs.

Abdomen large, ovate, boldly projecting over base of cephalo-thorax, grey, with dark brown markings, and a large brown patch at the centre; sides grey; at the highest point of the anterior extremity there is a recurved row of nine tubercles, the central one of which is by far the largest and most prominent; besides these there are on each side of the superior surface of the abdomen twelve tubercles, the first nine of which are distributed over three slightly procurved rows of three each; the fourth row on each side consists of two each, and the twelvth tubercle is solitary; the total number of tubercles is 33; the median portion of the superior surface is, with the exception of two rather deep circular depressions, smooth; inferior surface yellow-brown with dark markings and moderately clothed with hoary hairs laterally, and yellowish pubescence in the median line.

Epigyne, a small tri-lobed tubercular eminence, arched in front, hollow within.

Hab. Cooktown.

The specimen herein described was collected by Mr. E. A. C. Olive, of Cooktown, and presented to the Trustees of the Australian Museum by Mr. P. de la Garde, R.N., Paymaster of H.M.S. "Waterwitch."

DESCRIPTION of a NEW BIVALVE, *LIMA ALATA*, from SANTA CRUZ.

By C. Hedley, Conchologist.

Shell large, thick and strong, colour uncertain, in outline but slightly oblique, subelliptical, sharply truncate above and roundly produced below, in transverse section moderately rounded, in breadth two-thirds the length. Hinge line long, a third of the length of the valve, nearly straight, set at about an angle of thirty-five degrees to the median line of the valve. The narrow and shallow ligamental area is overhung by the small, sharp, produced umbo. The anterior auricle is enormously developed, thickened, reflected, and sharply and strongly recurved; the posterior is straight and moderately developed. The exterior is everywhere regularly furrowed by numerous small grooves equal to their interstices. These are about a hundred in number, are deepest and broadest on the auricles and above, finest on the centre of the valve. They diverge at a small angle on each side of a median line; transversely concentric growth lines indent and distort them.

Length, 70 mm.; breadth 50 mm.

Locality.—Santa Cruz, S. Pacific.

Type.—Australian Museum.

The material for the preceding description is a single, rather worn and discolored right valve, collected by Mr. J. Jennings on

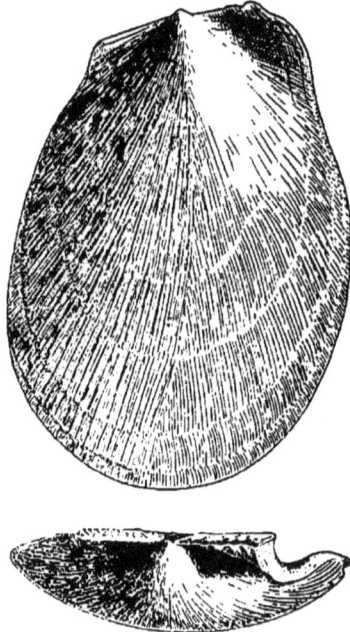

the beach of Santa Cruz Island in 1897, and presented by him to the Trustees. So unlike any other described species does it appear that I have considered that even such imperfect material should be utilised.

The brothers Adams, in grouping the recent species of *Lima*, set apart under the subgenus *Ctenoides* of Klein, *L. scabra*, Born, and *L. tenera*, Chemnitz, distinguished, among other features, by a sculpture of "ribs divaricate, meeting in the centre." Our novelty appears to find its nearest relations with these. More recently described species embraced by this character are :—*L. albicoma*, Dall., *L. concentrica*, Sowerby, and *L. murrayi*, Smith. *L. alata* appears to be longer than any of its immediate kin, with which I am not autopically acquainted, and to differ from any recent *Lima* in the development and sinuation of the anterior auricle.

<hr>

ORNITHOLOGICAL NOTES.

By ALFRED J. NORTH, C.M.Z.S., Ornithologist.

<hr>

IV.—ON A SPECIES OF PIGEON FREQUENTING THE ATOLLS OF THE ELLICE GROUP.

For nearly a quarter of a century various observers and writers have made reference to a species of Pigeon frequenting the Ellice Group. As I pointed out in my brief notes on the birds brought back from Funafuti by Mr. Hedley,[*] I could find no record of adult specimens having been obtained, but there was little doubt that the birds seen by Mr. Jansen on Funafuti in 1876, and by Mr. C. R. Swayne on Niu in 1895, were correctly identified by them as *Globicera pacifica*.

Prior to the departure of the Coral-Boring Expedition for Funafuti last year, I urged the desirability of procuring one or

<hr>

[*] Austr. Mus. Mem. III.—Atoll of Funafuti, pt. 1, Aves, 1896, p. 86.

more of these Pigeons if possible, and subsequently Mr. W. G.
Woolnough, B.Sc., succeeded in shooting the specimen herein
referred to, on the islet of Fualifeke, Funafuti Atoll, on 9th
August, 1897. The Pigeon was presented to the Trustees of the
Australian Museum. It proved to be a female *Globicera pacifica*,
slightly smaller, but precisely similar in plumage to an example
obtained about the same time from the Rev. H. A. Robertson of
Erromanga, New Hebrides. There is, however, scarcely any
indication of the knob at the base of the bill and it is probably
similar to the specimen on which Bonaparte bestowed the name
of *Globicera microcera*, a synonym of this species. This may be
due to immaturity, for the feathers surrounding the bill and on
the chin, are not quite so white as in examples obtained in other
islands of the Pacific, and in which the knob on the bill is more
developed. In the Ornithology of the "United States Exploring
Expedition,"[*] Cassin, writes as follows of this species: "The
knob at the base of the bill in this bird is not so greatly
developed as in some other species, at least this is the case in the
specimens now before us, from the collection of the Expedition.
This appendage occurs in several species of the group of fruit-
eating Pigeons, and is we suspect, not only peculiar to, or largest
in the male bird, but also most observable at the commencement
of the season of incubation, like the appendages on the head and
neck, or wattles, of the turkey."

The specimen of *G. pacifica* from Funafuti, measures :—total
length 12·5 inches, wing 8·8, tail 5·2, bill 0·95, tarsus, 0·95. A
female from Erromanga measures : –total length 14·5, wing 9·2,
tail, 5·6, bill imperf., tarsus 1.

Since the receipt of the above specimen, Mr. J. Stanley
Gardiner, B.A., of Caius College, Cambridge, has kindly sent me
a reprint from The Ibis for January 1898, containing an account
by Dr. Hans Gadow, of the birds collected by Mr. Gardiner on
Funafuti, and later on at Rotumah. Nine species were obtained
on the former atoll, of which *Numenius tahitensis*, *Charadrius
fulvus*, *Strepsilas interpres*, and *Gygis candida*, are additions to
its avi-fauna. Dr. Gadow also remarks "The following species
was observed but not obtained :—*Carpophaga pistrinaria.*"

Evidently this is the Pigeon Mr. Gardiner informed me, on
his return to Sydney, that he had seen on Funafuti, but was
unable to identify in the Museum. Hitherto, this species, of
which we have a fine series in the collection, has only been recorded
from the Solomon Islands, about a thousand miles from Funafuti.
It is a larger and much lighter coloured bird, and if well seen
could hardly be mistaken for *Globicera pacifica*. If Mr. Gardiner
is correct in his determination, there are two species of Pigeons
frequenting the Ellice Group.

---

[*] Cassin—U. S. Expl. Exped. Orn. p. 265 (1858).

The following is a list of the birds recorded up to date, that frequent these atolls, brought so much into prominence by the recent coral-boring expeditions :—

Urodynamis taitensis	Strepsilas interpres
Fregata aquila	Limosa novae-zealandiae
Demiegretta sacra	Anous stolidus
Globicera pacifica	Micranous leucocapillus
*Carpophaga pistrinaria	Procelsterna cerulea
Charadrius fulvus	Sterna anaestheta
Totanus incanus	Sterna melanauchen
Numenius tahitensis	Gygis candida

V.—ON THE OCCURRENCE OF *BUTASTUR TEESA* IN AUSTRALIA.

Some time ago Mr. Richard Grant of Lithgow, presented a skin of *Butastur teesa* to the Trustees, accompanied by the following note : "With regard to this Hawk, I shot it in a ring-barked tree, near the Bowenfels road, Lithgow. I do not know the exact date, but as near as I can remember it was in November 1889. I skinned it, also some Brown Hawks, that I shot the same day, and partly filled the skins out and put them away. I took no further notice of them until my brother returned home and drew my attention to this bird's plumage." Lithgow is situated in a valley of the Blue Mountains, 3007 feet above the level of the sea, and 96 miles west of Sydney. *Butastur teesa*, the White-eyed Buzzard Eagle is very abundant in some parts of India, which is the habitat of this species, but I can find no record of its having been obtained on any of the islands lying between India and Australia. *B. liventer*, which occurs in Java and Timor, or *B. indicus*, inhabiting Borneo, Sumatra, and the Phillipines, I should not have been so much surprised at obtaining on the Australian continent. The specimen of *B. teesa* procured at Lithgow, is similar to others in the collection from India, except in showing very little trace of the white mottlings on the wing-coverts. It is not in full adult plumage, for the sides of the throat and the spots on the breast are white instead of yellowish-white, otherwise it agrees with the description of the adult female given by Dr. Sharpe† in the "Catalogue of Birds in the British Museum."

VI.—ON A LIVING EXAMPLE OF *PSEPHOTUS CHRY-SOPTERYGIUS.*

Regarding this species, Gould, who described it, writes in his Handbook to the Birds of Australia,‡ as follows :—"One of the

---

* On the authority of Mr. Gardiner.
† Sharpe—Cat. Bds. Brit. Mus., i., p. 295, (1874).
‡ Gould, Handbk. Bds. Aust., ii., p. 65 (1865).

greatest pleasures enjoyed by the late celebrated botanist Robert
Brown, during the last thirty years of his life, was now and then
to show me a drawing of a Parrakeet made by one of the brothers
Bauer, from a specimen procured somewhere on the north coast
of Australia, but of which no specimen was preserved at the time,
and none had been sent to England, until several were brought
home by Mr. Elsey, a year or two prior to Mr. Brown's death.
On comparing these with the drawing made a least forty years
before, no doubt remained on my mind as to its having been made
from an example of this species. This, then, is one of the novelties
for which we are indebted to the explorations of A. C. Gregory Esq.
and I trust it may not be the last I shall have to characterize
through the researches of this intrepid traveller. Mr. Elsey, who,
as is well known, accompanied the expedition to the Victoria
River, obtained three examples— a male, a female, and a young bird
—all of which are now in our national collection. In the notes accompanying the specimens, Mr. Elsey states that they were procured on
the 14th of September 1856, in Lat. 18° S. and Long. 141° 33' E.,
and that their crops contained some monocotyledonous seeds."
    Since the above passage was written by Gould, so far as I am
aware, no additional information has been recorded of *Psephotus
chrysopterygius*, the rarest of all our Australian Parrakeets, and
the three specimens in the British Museum obtained by Mr. Elsey
in 1856 were the only ones known. It was therefore with extreme
pleasure that when passing one of the bird dealer's shops near
Circular Quay, in November 1897, my attention was arrested by
a living specimen of the Golden-shouldered Parrakeet, the first I
had seen, and previously known to me only by Gould's description
and figure. On making inquiries I found that it had been caught
by a bird-catcher in his nets about three months before in the
neighbourhood of Port Darwin, in the Northern Territory, and
was the only specimen that he had ever seen. Subsequently it
was acquired by the Trustees, and has since enlivened my room
with its cheerful notes. It bears confinement well and is exceedingly tame, except to strangers, feeding entirely on millet seed
and leaving untouched the canary seed with which it is mixed.
Like other members of this genus—which I have seen wade into
water to quench their thirst—it partakes freely of water. One
note of this species repeated several times at intervals of a second
apart is exceedingly sharp and shrill, and resembles the metallic
sound produced by quickly turning an unoiled key in a new and
close fitting lock. The remainder of its notes which are continued
for some time, is like the warbling of the Grass Parrakeet,
Melopsittacus undulatus, only much louder. This specimen
measures ten inches, and from the brilliancy of its plumage is
evidently an adult male. Gould's central figure of the male of
this species in his " Supplement to the Birds of Australia,"* is too

---

* Gould—Suppl. Bds. Austr., pl. 64 (1869).

large and robust ; a better idea of its size is conveyed by the upper figure of the supposed female. The figure of the male is fairly accurate in colour, except in the bill, cere and feet, which at all times it is a difficult matter to faithfully depict from dried skins. In the living example now before me a narrow line of turquoise blue separates the pale-yellow feathers of the forehead from the crown of the head, and the black feathers of the latter extend in a central stripe on to the nape ; the bill is horn-white, faintly shaded with bluish-grey at the base, and the cere, legs, feet and *claws* are of a pale cinnabar-flesh colour.

It is worthy of remark, that forty years elapsed between Bauer making a drawing of this bird, and Elsey obtaining the first specimens, and that nearly a half century has since passed away before the discovery of another specimen. Only four examples and a drawing of this bird during a period of eighty-two years, fully entitle it to the distinction of being the rarest of all our Australian Parrakeets.

Addendum.—Since the above was in type I have received Part iv. of the Proceedings of the Zoological Society of London, for 1898, and find in the list of additions to the Gardens, that a pair of these birds was purchased by the Society on the 10th of March, 1897.

VII.—On the EXTENSION of the RANGE of *PHÆTON CANDIDUS* to NEW SOUTH WALES and LORD HOWE ISLAND.

Climatic influences are among the most important factors in the distribution of species, and the recent heavy easterly gales of February 10th, 11th, 12th of the present year, which caused so much disaster to the shipping on the coast of New South Wales, have been the means of increasing the number of birds included in its avifauna. On the 15th of February an immature specimen of *Phæton candidus*, in the flesh, was presented to the Trustees by Mr. Henry Burns, who had picked it up in a dying condition, the previous day, on the shores of Botany Bay. This species was not met with by Gould, neither is it mentioned in any of his works on Australian birds. Dr. E. P. Ramsay has, however, in his "Tabular List of Australian Birds" included Cape York and Wide Bay, among the numerous localities over which it enjoys a range. Previously it was not represented in the Museum by an Australian specimen, but there is portion of a skin in a slightly advanced stage of immaturity from Lord Howe Island, obtained there by Mr. D. Love in May 1890 ; another new locality for this species. This wanderer over the intertropical zone of the Atlantic, Indian, and Pacific Oceans, has been recorded, among other localities, by Count Salvadori in his " Ornitologia della Papausia e delle Molucche" from Florida, Cuba, Costa Rica, Jamaica, the Bermudas,

Madagascar, Mauritius, Bourbon, Rodriguez, the Seychelles, India, Ceylon, Andaman Islands, Solomon and Friendly Islands, New Caledonia, the Marquesas, Gilberts, Marshall, and Pelew Islands.

The immature specimen of *P. candidus*, obtained at Botany is silky-white with the upper parts beautifully marked with black crescentic and arrow-headed cross-bars similar to immature examples of *P. rubricauda*. Total length in the flesh exclusive of the two central tail-feathers which are imperfect, 14·5 inches, wing 10·1, tail-feathers next the central pair 4·4, bill 1·8, tarsus 0·7; bill and legs ashy-flesh colour, feet black.

An adult mounted specimen received from America is silky-white, with a crescent in front and a line behind the eye, a stripe along the wing-coverts terminating on the innermost secondaries and scapulars, and a band on the first four primaries and some of the elongated flank-feathers, black ; shafts of the wing and tail-feathers, except at the tips, black. Total length 22 inches, wing 10·5, central tail-feathers 11, bill 1·8, tarsus 0·72 ; bill and legs pale yellow, feet black.

Dr. J. C. Cox informs me that immediately after the same storm in February 1898, in company with Major Ferguson, he observed a specimen of *Pelagodroma marina* between Cockatoo Island and Iron Cove, skipping over the surface of the water and following in the wake of their boat. Living examples of this species are extremely rare in New South Wales waters, and are only seen after unusually severe easterly or southerly gales.

ADDENDA to CATALOGUE of AUSTRALIAN METEORITES.[*]

By T. COOKSEY, Ph.D., B.Sc., Mineralogist.

BINGARA.—Prof. Liversidge wishes it to be stated that he is not in possession of any portion of the Bingara Meteorite.

YARDEA—The following additional information has been received from Dr. E. C. Stirling :—

Type.—Siderite.

Weight.—7 lbs. 3½ozs.

Locality.—Four miles S. of Yardea Station, Gawler Ranges, South Australia.

Finder and Date.—Found in 1875.

Coll.—The Museum, Public Library, Museum and Art Gallery of S.A., Adelaide.

[*] Rec. Aust. Mus. III., 3, p. 55 - 60.

DESCRIPTION OF A RING-TAILED OPOSSUM,
REGARDED AS A VARIETY OF
PSEUDOCHIRUS HERBERTENSIS, COLLETT.

By EDGAR R. WAITE, F.L.S., Zoologist.

HAVING recently had occasion to overhaul our duplicate collection of Marsupials, my attention was arrested by a number of specimens labelled *Pseudochirus herbertensis*. Out of twenty-two examples, seventeen were undoubtedly of this species, but the remainder presented some differences. I therefore consulted Mr. Robert Grant, by whom the Herbert River animals were procured. Unhesitatingly picking out the five forms I had marked, he told me that he was convinced that they were quite distinct from the "Outas" (*P. herbertensis*), and had so reported when unpacking the collection in 1889.

The following notes are supplied to me by Mr. Grant :— Although found in the same district as *P. herbertensis* and *P. lemuroides,* the smaller and much rarer animal was obtained within a comparatively limited area, the exact locality being known as the Boar Pocket, on the Tinaroo track, near Cairns, Queensland (or, in the language of the blacks, " Wamaranna Rigarami"). This spot lies low, and is swampy. *P. herbertensis,* although obtained on the outskirts of the swamp, is an inhabitant of the higher land and ascends to the tops of tallest trees, while the swamp animal is not a high climber, and several of them were found in the Davidsonia plum tree (*D. pruriens,* F. v. M.)

Another interesting fact is that this animal builds a nest or drey not unlike that of the common Ring-tailed Opossum (*P. peregrinus*), but more ball-like in shape. In passing it may be mentioned that this latter species generally builds its nest near to creeks or in moist gullies. *P. herbertensis* never builds a nest, so the natives say.

The blacks instantly recognised the animal as distinct and called it " Moki poki." Mr. Grant would throw upon the ground a few examples of *P. herbertensis,* which the blacks would at once name " Outa." One of the swamp forms would next be cast down, when they would laugh and remark, " No more ' Outa,'—' Moki poki.' "

Although smaller than the " Outa," the animals obtained were fully adult, as our collector took young ones from the pouches of the females.

A

For reasons hereafter given, I do not feel justified in regarding this animal as a distinct species, although in view of its peculiarities of fur, tail, and habit, it may ultimately be deemed worthy of specific rank. At present I prefer to regard it as a well-marked variety of *P. herbertensis*, and deserving of at least a varietal name. I have, therefore, much pleasure in associating with it the name of Dr. Robert Collett, of Christiania, whose researches into the fauna of Australia, and of the genus *Pseudochirus* in particular, are well known.

This form will therefore be known as :—

PSEUDOCHIRUS HERBERTENSIS, var. COLLETTI.

Animal smaller. Fur markedly longer, less wavy, much finer and softer to the touch than in the typical form. Much greyer and darker in colour, the hairs behind the shoulders being usually tipped with white or pale yellow ; the rump and coloured portion of the tail black. The ears are rich rufous without and the chin is grey. The naked portion beneath the tail is smooth, not sharply defined from the hairy part and of less extent than in the typical race.

The dimensions of five animals are as follows :—

	A.	B.	C.	D.	E.
Head and body	320	320	300	290	260 mm.
Tail	310	310	290	280	245 mm.

Skull.—Excepting for its relatively smaller size, the skull scarcely differs from typical examples, and mainly for this reason I hesitate to accord the form more than varietal rank. It may be noticed, however, that the facial index is higher than that of *P. herbertensis*, as determined by Thomas.[*]

The principal skull dimensions are as follows :—

	mm.
Basal length	62·4
Greatest breadth	35·6
Nasals, length	23·8
,, greatest breadth	8·5
,, least breadth	3·7
Constriction, breadth	7·4
Palate, length	37·9
,, breadth outside M^2	18·2
,, ,, inside M^2	12·1
Palatal foramen	5·4
Basi-cranial axis	21·2
Basi-facial axis	41·0
Facial index	193

---

[*] Thomas—Brit. Mus. Cat. Marsup., 1888, p. 185.

				mm.
Teeth,	horizontal length of I^2	...	...	1·5
,,	height of *Canine* ...	...	...	2·3
,,	length of P^1	...	...	3·2
,,	length of $M^{1·3}$	..	...	11·4
,,	diastema of I^3 and *C.*	...	...	3·7
,,	,, *C.* and P^1	...	...	2·8
,,	,, P^1 and P^3	...	...	1·0
,,	length of lower I^1 ...	...	...	9·8

It is interesting to notice that the structure of the tail corresponds with the habits of the animals ; thus, in *P. herbertensis*, which ascends the highest trees, the lower surface is naked for a greater portion of its length and is roughened so as to afford a secure grip of the topmost wind-swayed branches. In its more lowly habit, *P. colletti* avoids such positions, and has therefore less need of special adaptation.

It may be mentioned that *P. mongon*, De Vis,* of which we hold co-types from the describer, exhibits none of the characters here sought to be emphasised, and except in the markings does not differ from typical examples of *P. herbertensis*, as previously determined.

THE NEST OR DREY OF THE RING-TAILED OPOSSUM, (*PSEUDOCHIRUS PEREGRINUS*, BODD).

By EDGAR R. WAITE, F.L.S., Zoologist.

(Plate xix.)

ONE of the most peculiar and interesting habits of the Ring-tailed Opossum (*Pseudochirus peregrinus*, Bodd), is that of making a nest or drey. Although well-known, but little appears to have been written on the subject beyond the notice that it is not unlike that of the European Squirrel.

I as often found the drey of this latter animal in a hole in a tree as among the branches, a situation never utilised by the Opossum. The nest of the Marsupial may be constructed either in a fork or upon a platform of interlaced twigs. A thick bush is more favoured than a tree, but almost any growth, if sufficiently dense, may be made use of : the Lilly Pilly (*Eugenia*), offers a congenial retreat, as does also the Tea Tree (*Melaleuca*), its long strips of loose bark being frequently woven into the nest. The native "Oaks" (*Casuarina*), and the Wattles (*Acacia*), are further favourites. Preference is shown for the neighbourhood of water.

---

* De Vis—Proc. Linn. Soc. N.S. Wales (2) i., 1887, p. 1130.

The animal is usually a lowly builder, especially if a thick bush be selected, the nest may then be within seven or eight feet of the ground. In more open foliage it may be raised twenty or even thirty feet : in such cases a more careful attempt is made to conceal the structure, hanging bark, drooping moss, or a mass of foliage being pressed into service.

The nest is carefully and neatly made. It has for foundation small sticks from which finer twigs are carried upwards and brought together in the form of an elongated dome. An aperture is provided at one end through which the animal gains access ; thus giving a bottle-like appearance to the fabrication. Leaves, grasses, and mosses are skilfully woven into the structure, the mouth excepted ; this is narrowed and projects somewhat, and is usually composed of naked twigs alone. The interior is smoothly lined with fine grasses, and the whole forms a compact and fairly firm structure.

We have recently received from Mr. J. M. Cantle a drey of this Opossum, wherein the usual type has been considerably departed from. Grasses and mosses have been entirely discarded, and their place supplied by the fronds of ferns. Ferns only are to be seen within, but I am inclined to think that the drey was not quite completed and that a smoother lining would have been provided. Mr. Thomas Whitelegge has identified the fronds as wholly of *Pteris esculenta*, Forst. This example measures:—Length, 14 inches ; breadth, 11 inches ; greatest circumference, 36 inches.

The drey is occupied only by the female ; but whether the male also takes part in its construction, I have been unable to learn. If any of my readers have made observations on this or any other matter connected with the animal, I shall be pleased to hear from them.

Although one usually associates the drey as a habitation of the female while carrying young in her pouch, it would appear, as I learn from one or two sources, that the young are occasionally left in the drey during the absence of the mother, and Mr. R. Grant tells me that he once kept alive, two young Opossums which he obtained under such circumstances.

If disturbed in her retreat, the mother may become very savage. On one occasion Mr. J. A. Thorpe having too incautiously sought to investigate the nest, received very severe wounds—the animal making its teeth meet in the fleshy part of the hand. This nest was placed in a small Eucalypt, and was reached from the ground by bending down the branches.

It remains to be mentioned that the Ring-tailed Opossum does not enjoy a monopoly in nest making among members of the genus *Pseudochirus*. As described elsewhere in this publication (p. 91), such a habit is practised by *P. herbertensis*, var. *colletti*, the nest differing only, as far as we know, in being more ball-like in shape.

OBSERVATIONS on *TESTUDO NIGRITA*, DUM. & BIBR.

By EDGAR R. WAITE, F.L.S., Zoologist.

(Plates xx., xxi., and xxii.)

UNTIL recently, living in the grounds of the Hospital at Glades-ville, near Sydney, were two Gigantic Land Tortoises. In April, 1896, I availed myself of a long-standing invitation from Dr. Eric Sinclair to inspect these tortoises with a view of determining the species and to verify, if possible, the common belief that the larger of the two originally came from the Galapagos Islands, the home of one of the three races of gigantic land tortoises.

I then understood that little or nothing was known of the history of these huge Chelonians; it was therefore, in the first place, necessary to determine the species. An examination showed that although the two individuals differed slightly, attributable to their being of opposite sex, they were of the same species. The absence of a nuchal plate, together with a divided gular, and the presence of an enlarged scute on the inner side of the fore-limb, at once indicated that they were from the Galapagos Islands. Of the six species inhabiting the group, three only have the shields of the carapace concentrically striated, as exhibited by our specimens. The anterior declivity of the carapace, taken in con-junction with the feature of the plastron, namely, being deeply notched behind, at once enables us to determine the species as :—

TESTUDO NIGRITA, *Dum. & Bib.**

I had hoped to fully describe the species, and so close a gap in our knowledge of the several forms ; pressure of work has hitherto prevented my doing so, and, notwithstanding the facilities we possess in the way of material, I cannot look forward to sufficient leisure at any near date. The male is now in London, but is still, I believe, alive, so that its osteological characters are not ascertainable. The female died in August, 1896, and then passed into the possession of the Museum. Being desirous of preserving it in a life-like condition, and at the same time not wishing to sacrifice such a valuable skeleton, a novel experiment was tried and proved to be most successful. Casts were taken of the carapace, plastron, and head ; the skin was next carefully removed, and so worked up in conjunction with the casts that no one could now detect the deception, and the production occupies a

---

* Duméril et Bibron—Erpét. Génér., ii., 1835, p. 80.

prominent position in the Reptile Gallery. The entire skeleton was thus preserved, and being carefully articulated, is exhibited in the Osteological Gallery. It is, therefore, still available for study should opportunity occur. (See Plate xxii.)

Failing the work I had proposed, it appears to be advisable to publish such information as I have been able to glean relative to the history of the specimens. In this connection I have to thank Dr. Sinclair, who has spared no pains, and who has himself made some of the observations below recorded.

In the Hospital grounds these reptiles had almost unlimited freedom, and their feed of herbage was largely supplemented by vegetables—lettuces being much appreciated; an entire plant was taken from the hand, and after two or three movements of the jaws was swallowed. Eating seemed to be a constant occupation. For my benefit the tortoises were prodded about the grounds; they were very disinclined to move and only did so in response to repeated persuasions from a stick directed at the hind limbs. I was much interested to notice that when touched in this way the tortoise would suddenly drop its shell over the leg prodded, and so endeavour to protect itself, and at the same time cripple the offender. One of the men engaged in trundling the reptile used his foot for the purpose, but this was considered an unwise proceeding, for on one occasion, I was told, when kicking the leg of the reptile an attendant had his foot badly crushed.

When turning the smaller reptile (the female) about, two or three men proved to be sufficient. When the male was under consideration, Dr. Sinclair told off five men, and in consequence of the enormous weight and the struggles of the huge creature, they were scarcely able to turn it over. More help was required when it had to be placed on the weighing machine, and even when once there it managed, by hitching its claws into the standards, to force itself off, despite all efforts to prevent it. Finally, the correct weight was obtained, and when the reptile was permitted to regain its legs I noticed that blood was issuing from between the shields of the carapace.

The tortoises were propped up so as to show their under-surfaces. The photographs I then took are reproduced on Plate xxi., and exhibit several points of interest, not otherwise observable. That of the female (Fig. 2), propped against the Hospital wall, shows the enlarged scute on the fore-leg very clearly, also the depression in the plastron, the divided gular and the notch between the xiphiplastra with the angles rounded; the short tail is also well illustrated, as are also the sutures of the plastron and the characteristically wrinkled skin.

Owing to its weight and bulk, the male could not be retained vertically against the wall, and as it absolutely refused to be further handled, we had to make the best of things. Although not so good, the photograph (Plate xxi., Fig. 1) shows sufficiently the principal secondary sexual characters ; these are :—the great concavity of the plastron, the angular aspect of the xiphiplastral plates, and the comparatively long tail.

At the time of my visit, the upper posterior portion of the carapace of the female was much abraided, owing to the efforts of the male, who had been paying her considerable attention.

As each tortoise has its individual history, the two may now be treated separately.

THE MALE TORTOISE.

The male was commonly called " Rotumah " in the gardens and is, I find, directly traceable to the Galapagos Group, but from which island it was obtained is not known. This is to be regretted, as it is the only species whose definite habitat has not been ascertained.

For particulars of the early history of this tortoise I am indebted to Miss Annie C. E. MacDonald. About the year 1866 it was given to her father, the late Alexander MacDonald, by King George of Tonga, and was what was called a " chief's gift," that is, a gift supposed to pass between two great chiefs of equal standing. When taken to Tonga from Rotumah, the reptile caused a great sensation among the natives, and was presented to Mr. MacDonald in recognition of his kindness to the King's son when in Sydney, both father and son having taken a violent fancy to the well-known trader.

The tortoise was brought to Sydney in the schooner " Ida," one of MacDonald and Smith's whalers. Captain Howard, who was in command of the vessel, had known the tortoise for fifty years previously on the island of Rotumah, it having been landed there from the Galapagos Islands by an American whaler many years before. It was within the memory of the inhabitants, always of the same size.

From 1866 to the end of 1896 the tortoise lived in Sydney, and at the later date was removed to England, having been purchased, I understand, by the Hon. Walter Rothschild for his menagerie at Tring. When the tortoise passed into the possession of Mr. MacDonald he had it photographed, and the accompanying illustration (Plate xx.) is reproduced from a copy kindly lent by Miss MacDonald. On the margin of this copy are a number of measurements made by the owner at the time. These I reproduce (in lit.) below :—

		Ft.	in.
Length, nose to tail		6	2
„	shell ...	4	7½
„	across shell	5	10½
„	under shell	3	0
Girth	... ...	8	3

	Ft.	in.
Height, lying down	2	$2\frac{1}{4}$
„ standing up	3	1
Front leg, under knee...	1	$7\frac{1}{2}$
„ round elbow	2	$1\frac{1}{2}$
„ round foot	1	11
Hind leg, instep	1	$9\frac{1}{2}$
Head, round head	1	7
Weight—5 cwt. 2 qrs. 26 lbs. = 642 lbs.		

These figures were, I find, published in 1884 by Dr. J. C. Cox,[*] who, while stating that the reptile came from Rotumah, was unaware of its having been at Tonga prior to passing into the possession of Mr. MacDonald. From this account it appears that the Galapagos Group was then thought to be the original habitat of both the tortoises, for Dr. Cox writes:—" I am not at all sure as to what species these two Tortoises belong, but they are supposed to come from Galapagos Archipelago."

The next measurements available, are a few made by Dr. Sinclair, in 1893, for comparison with the gigantic tortoise living at Port Louis, in Mauritius, of which an account appeared in the *Illustrated London News* :—

	Port Louis.		"Rotumah."	
	Ft.	in.	Ft.	in.
Fore leg...	1	$7\frac{3}{4}$ ...	1	9
Hind leg	1	0 ...	1	5
Circumference (girth)	7	0 ...	8	4
„ round shell at base	—	...	11	6
Neck and head...	1	$3\frac{1}{2}$ ...	1	5
Head	0	7 ...	0	7

On 23rd April, 1896, I examined the tortoise, and took the following dimensions, which, for comparison with the foregoing, are reduced to English scale :—

	Carapace.	mm.		Ft.	in.
Length over curve		1420	=	4	8
„ in straight line		1284	=	4	$2\frac{1}{2}$
Width over curve		1680	=	5	6
„ in straight line		890	=	2	11
	Plastron.				
Length		980	=	3	$2\frac{1}{2}$
Width		915	=	3	0
Depth of concavity		130	=	0	$5\frac{1}{8}$
	Caudal plate.				
Length		133	=	0	$5\frac{1}{4}$
Width		250	=	0	$9\frac{3}{4}$
	Weight, 575 lbs.				

---

[*] Cox—Proc. Linn. Soc. N.S. Wales, viii, 1884, p. 531.
[†] *Illust. Lond. News*, 3rd Dec., 1892, p. 715.

The Female Tortoise.

As I learn from Dr. Sinclair, the female was brought to Sydney in 1853 by the American Whaler "Winslow." It was then but a baby and weighed 56 lbs.* No further observations appear to have been made until 1884, when the following figures were published by Dr. Cox (*loc. cit.*):—

					Ft.	in.
Length, nose to tail (" no tail ")?	...	...	...	...	5	10½
„ shell	...	...	...	...	4	0
„ across shell	...	..	...	...	5	0
„ under shell	...	...	...	...	2	4½
Girth	...	...	...	...	6	7½
Front leg, round elbow	...	...	...	...	1	5
Head, round	...	...	...	...	1	3

In 1893 its weight was ascertained to be 368 lbs. About this date it was placed in a paddock with the male, and in September, 1895, it was found to have deposited six eggs in a rubbish heap. These eggs were at once forwarded to the Museum, when I took the dimensions below recorded. The following measurements of the female were made by me in April, 1896. In August of the same year this tortoise died, and as already stated, was forwarded to the Museum. The ovaries were in an enlarged condition, and it seems probable that had she lived, the tortoise would have again produced eggs.

	Carapace.			mm.		Ft.	in.
Length over curve	...	...	...	1195	=	3	11
„ in straight line		...	...	915	=	3	0
Width over curve	...	...	...	1295	=	4	3
„ in straight line		...	...	740	=	2	5

	Plastron.						
Length ...	...	...	...	775	=	2	6½
Width ...	...	...	...	740	=	2	5
Depth of concavity		...	...	43	=	0	1¾

	Caudal plate.						
Length ...	...	...	...	115	=	0	4½
Width ...	...	...	...	185	=	0	7¼

Weight, 320½ lbs.

The depression in the plastron, although much less than in the male, is yet very noticeable; a feature scarcely realised by Dr. Günther when writing his Monograph,† for he regarded the type

---

* See also—The Curator's Annual Report for 1897: Report of the Trustees for 1897, p. 6.

† Günther—Gigantic Land Tortoises in the British Museum, 1877, p. 71.

of the species as a young male "inasmuch as the sternum shows
a slight concavity." It will be noticed that between 1893 and
1896, the female had lost weight to the extent of about 47 lbs.,
and that in thirty years the male had lost 67 lbs. It is to be
regretted that such little data is available from which to draw
deductions. There is no doubt, however, that under Mr.
Rothschild's care, future developments of the male will be
carefully recorded.

<h2 style="text-align:center">EGGS.</h2>

Porter* (fide Günther) writing on the tortoises of the Galapagos
Islands, remarks :—"The eggs are perfectly round, white, and of
2½ inches diameter." A glance at the following table shows that
the eggs of *T. nigrita* are not far removed from the spherical, the
accompanying cut representing the shape of the egg marked
F. Reduced to inches, the longest diameter of the largest example
is barely 2¾. Darwin† measured an egg having a circumference
of 7¾ inches. Specimen C measures 7¼ inches in circumference.

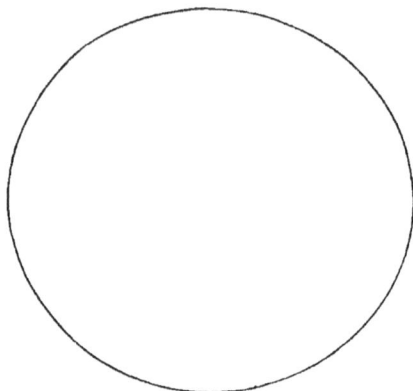

The egg marked F is the largest, but its weight was dispropor-
tionate and actually less than the others. On emptying this egg,
the contents were found to be abnormal. It has a very rough
surface; all the others are tolerably smooth. Specimens B and
E were returned to Dr. Sinclair; the others are in the Museum
collection.

---

* Porter—Journal of a Cruise made to the Pacific Ocean, New York,
1822, pp. 215, 216.

† Darwin—Voyage of the "Beagle," iii., 1839, p. 464.

EGG.	WEIGHT IN GRAMMES.	DIMENSIONS IN MILLIMETRES.	GREATEST CIRCUMFERENCE.
A.	99·2	56·9 × 53·4	175·2
B.	106·7	56·8 × 55·5	— —
C.	109·8	57·8 × 55·5	181·0
D.	105·0	58·2 × 55·2	181·8
E.	106·4	58·9 × 55·0	—
F.	85·8	59·8 × 54·8	180·4

OTHER TORTOISES.

In the Museum collection are four examples of *Testudo nigrita*. These may be tabulated as follows :—

A. The female (and skeleton) already dealt with.
B. A smaller female, mounted in the Museum.
C. Carapace and plastron of another example, possibly a male.
D. Skeleton of a much larger specimen, also believed to be a male.

The last named is the only other one of which I can gather any information. It was brought to Sydney in a schooner about the year 1860, and from that time to the date of its death it was an inhabitant of the Museum grounds. It not infrequently wandered into the street, and on one such occasion a cart collided with it and broke in part of the carapace, the injury being now very apparent. The specimen has not been well cared for in the past, and the mandible and portion of the tail bones have been lost. This collection of four is an interesting one, and shows well to what extent the shape and structure of the shell alters with age. In the younger examples the concentric striæ of the shields are extremely well marked (almost absolute in the adult), the free edges of the carapace are much more deflected, especially on the posterior border, while the costal sutures are more deeply cleft. Another feature exhibited by the immature form is the greater prominence of the knobby protuberances of the carapace.

The dimensions of the three smaller examples have been ascertained, as below :—

	B ♀	C ♂	D ♂
Carapace.	mm.	mm.	mm.
Length over curve ...	905	840	930
„ in straight line	710	715	755
Width over curve ...	950	920	960
„ in straight line	540	515	550
Plastron.			
Length	590	580	640
Width	590	510	540

	B ♀ mm.	C ♂ mm.	D ♂ mm.
Caudal plate.			
Length	75	75	85
Width	160	150	155

THE SKELETON.

Although I have not now the opportunity of carefully describing
and figuring the bones, some measurements have been attempted.
Such features have been selected as would be comparable with
those of other species of gigantic tortoises rendered by Günther.
As already intimated, two specimens are available, namely a larger
female (A) and a smaller male (D). It is necessary to mention
that, as both the skeletons are articulated, some of the measurements are approximate only :—

	A ♀ mm.	D ♂ mm.
Pectoral arch.—		
Scapula, length	220	165
,, least circumference ...	92	74
Glenoid cavity, greatest diameter...	55	48
Coracoid, length	120	86
,, greatest width	94	70
Precoracoid, length...	90	68
Fore limb.—		
Humerus, length	208	200
,, least circumference	110	96
,, head, longest diameter... ...	45	...
,, ,, shortest diameter ...	42	...
,, condyles, breadth	90	73
Ulna, least width	31·5	22
Radius, length	127	...
,, least circumference	64	48
Pelvic arch.—		
Pelvis, ilium to symphysis (vertical) ...	148	125
,, longitudinal diameter ... *153 – 178		125
,, distance between ilio-pubic pro- minences	107	80
,, foramen obturatorium	50	32
,, symphysial bridge, width	30	31
,, ,, ,, depth	28	20
,, ischium, least breadth of posterior part	69	63
,, ilium, length	158	110
,, ,, least breadth...	35	27

* See observation and figure on p. 103.

Hind limb.—	A ♀	D ♂
Femur, length	182	150
,, least circumference	99	76
,, head, longest diameter	59	46
,, condyles, breadth	81·5	60
Tibia, length	140	110
,, least circumference	99	76
Fibula, length	134	109
,, least circumference	48	39

In references to the pelvis, I fail to find mention of the epipubic cartilage becoming ossified. Such is distinctly the condition in the skeleton of the female, and the following sketch shows its form and relative position.

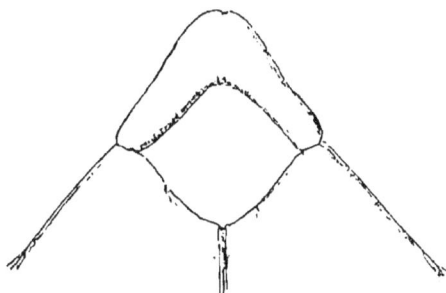

This bone is not ankylosed with the pubic, and may therefore be easily lost during the process of maceration. The additional figure in the foregoing table of measurements, is inclusive of this epipubic bone.

The apex of the entoplastral spine reaches a point situated 220 mm. behind the edge of the gular prominence, it has a width of 12 mm., and is quite free from the plastron for 45 mm. at its distal extremity.

The accompanying Plate (xxii.) is reproduced from a photograph of the skeleton of the female, from which one side has been removed to show the internal structure. It will be seen that the horny sheaths covering the jaw have been retained and that the hyoids are suspended in their natural position. The extent and attachment of the epipubic bone, above referred to, may also be traced.

NOTES ON SNAKES.

By EDGAR R. WAITE, F.L.S., Zoologist.

THE first two snakes below mentioned were included in a small collection made by the Rev. W. G. Ivens in the Solomon Islands. As both differ somewhat from the descriptions of the respective species, opportunity is taken to point out their individual peculiarities. The third note deals with an Australian species whose habitat has been the subject of some uncertainty.

I.—DENISONIA MELANURA, *Boul.*

Hoplocephalus melanurus, Proc. Zool. 1888, p. 88, and 1890, p. 30, pl. ii., fig. 1.

Denisonia melanura, Brit. Mus. Cat. Snakes (2nd ed.) iii., 1896, p. 345.

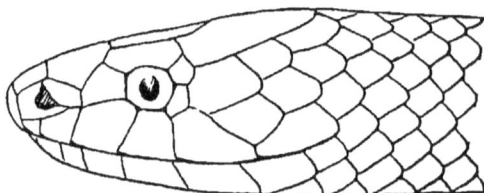

The most noticeable difference between our specimen of *Deni-sonia melanura* and those previously described, is to be found in the circumstance that it possesses six upper labials instead of seven. The lost plate occurs between the fifth and the ultimate labials, but fails to reach the mouth, as shown in the accompanying figure. This is perhaps an individual peculiarity and worthy of notice only as such. The frontal, however, is considerably longer than in the British Museum specimens, being as long as the prefrontals and internasals combined, and two-thirds the length of the parietals. The tail is not black as described, but similar in color to the body, likewise crossed by dark bands.

Having but a single example, I hesitate to create a new species. Pending further material, it may for the present be known as :—

Denisonia melanura, var. *boulengeri.*

The scale formula is as follows: Scales in 15 rows, ventrals 170 ; anal divided ; sub-caudals 43. Total length 850 mm.; tail 125 mm.

II.—MICROPECHIS ELAPOIDES, *Boul.*

Hoplocephalus elapoides, Proc. Zool. Soc. 1890, p. 30, pl. ii., fig. 3.
Micropechis elapoides, Brit. Mus. Cat. Snakes, (2nd ed.) iii., 1896,
p. 347.

The only particular in which this specimen differs from the type
is in the extent of its markings. Instead of the black bands being
less than twice the width of the interspaces they are five or
six times as wide. In Boulenger's figure the black bands are
shown as wide at the sides as on the vertebral line. In our
example they are so disposed that the cream ground colour appears
as a series of inverted Vs when viewed laterally. The condition
may be illustrated by supposing a number of flexible pennies were
laid upon the back within an eighth of an inch of each other and
then folded down the sides. The bands encircle the tail and the
ventral scales are generally edged and clouded with black. The
first black band commences close behind the parietals, which,
together with the snout and ocular region are also black.

III.—FURINA CALONOTA, *Dum. & Bibr.*

Furina calonotos, Erpét. Génér., 1854, vii., p. 1241, pl. lxxv. *b.*

The habitat of this species having always been in question, and
even yet doubtful, it is with satisfaction that I am able to remove
the uncertainty. The authors state that Verreaux sent two
examples from Tasmania in 1844. After a quarter of a century
a specimen was obtained by the British Museum, the locality
being given as Baranquilla at the mouth of the River Magdalena
in Columbia. Doubts were cast on this when *Brachyurophis
(Rhynchelaps)* purchased as from the same locality, was discovered
to be an Australian genus.

According to the British Museum Catalogue, the locality is
doubtfully West Australia, as explained by the footnote* :—

"The specimen was purchased as from Baranquilla, Columbia,
together with a specimen of the W. Australian *Rhynchelaps
semifasciatus.*"

Quite recently we received, by presentation from Mr. Henry
Richards, an interesting series of Reptiles from West Australia,
including a small example of *Furina calonota*, the locality being
rendered as Claremont, five miles from Perth.

* Brit. Mus. Cat. Snakes, (2nd ed.) iii., 1896, p. 407.

ORNITHOLOGICAL NOTES.

By Alfred J. North, C.M.Z.S., Ornithologist.

VIII.—DESCRIPTION of a NEW SPECIES of HONEY-EATER from NORTH QUEENSLAND.

Ptilotis leilavalensis, sp. nov.

Adult.—General colour above pale ashy-brown tinged with yellow, the upper tail-coverts more distinctly shaded with yellow; scapulars and upper wing-coverts like the back; quills brown, strongly washed with bright olive yellow, the apical portion of the outer webs of the primaries and the tips of the secondaries having whitish edges; tail-feathers brown with whitish tips, the two central feathers, and outer webs of the remainder washed with bright olive yellow; lores, forehead, sides of the head, cheeks and ear-coverts bright olive yellow ; behind the ear-coverts a patch of silky-white feathers; chin, throat, and fore neck pale olive-yellow passing into fawn-white, tinged with yellow on the breast and abdomen, lower portion of the abdomen and the under tail-coverts pale yellow ; bill black; legs and feet fleshy-brown. Total length of skin 6 inches, wing 2·8, tail 2·7, bill 0·42, tarsus 0·75.

Habitat. Fullerton River, Burke District, North Queensland.

Type. In the Australian Museum, Sydney.

Observations. This species, which will be vernacularly known as the Lesser White-plumed Honeyeater, is allied to *P. penicillata* and *P. flavescens.* From the former it may be distinguished principally by its smaller size, more brightly coloured head, and otherwise generally paler plumage, also by the absence of the blackish line which separates the silky-white patch of feathers from the ear-coverts. In size, and general colour of plumage, except the ear-coverts, it closely approaches *P. flavescens.*

The specimen from which the above description is taken was presented to the Trustees by Dr. W. Macgillivray of Hamilton, Victoria. It was obtained by his brother, Mr. A. S. Macgillivray, of Leilavale Station, Fullerton River, thirty miles east of Cloncurry Township, North Queensland, who states that these birds are fairly common in the Tea-trees along the river. The nest and eggs were secured at the same time, but the latter were unfortunately broken.

For the sake of comparison the following measurements are given

		Total length of skin.	Wing.	Tail.	Bill.	Tarsus.	Localities.
P. leilavalensis, ad. sk		6 inches	2·8	2·7	0·42	0·75	Fullerton River, N. Queensland.
P. penicillata ♂ ad. sk.		6·7 ,,	3·45	3·3	0·5	0·82	Near Adelaide, S. Australia.
,,	♂ ad. sk.	6·6 ,,	3·3	3·2	0·47	0·8	Bourke, Western N. S. Wales.
,,	♂ ad. sk.	6·5 ,,	3·3	3·1	0·47	0·8	Dawson River, Queensland.
P. flavescens,	♂ ad. sk.	5·7 ,,	2·95	2·7	0·5	0·7	Georgetown, Gulf District.
,,	♀ ad. sk.	5·4 ,,	2·92	2·7	0·48	0·7	Derby, N. West Australia.

IX.—DESCRIPTION OF THE NEST AND EGGS OF *MICRŒCA PALLIDA*, DE VIS.

Dr. W. Macgillivray has kindly forwarded the following description of the nest of this species, together with the eggs and skin of the bird for identification.

"Two nests of this species of *Micrœca* were taken by my brother Mr. A. S. Macgillivray on Leilavale Station, Fullerton River, North Queensland, between the 20th and 25th December, 1897. They were built rather low down on horizontal branches in a patch of Gidden scrub and each contained a pair of eggs. A nest my brother sent was slightly smaller but more substantially built than that of *M. fascinans*, and of much the same material, the outside being ornamented with bits of bark and lichen attached by means of cobweb."

The eggs are oval in form, one specimen having a pale bluish-grey ground colour, which is freckled and spotted with faint purple and purplish-brown, predominating and becoming darker as usual on the thicker end of the shell; the other is of a warm stone white ground colour, and in many places the markings which are of a light reddish-purple are confluent and intermingled with faint underlying spots of pale greyish-lilac. Length (A) 0·69 x 0·55 inch ; (B) 0·67 x 0·56 inch.

The range of this species probably extends right across the northern portion of the Australian Continent, for there are specimens in the Museum, obtained by Mr. E. J. Cairn at Derby, North-west Australia in 1886.

B

CONTRIBUTION to a KNOWLEDGE of PAPUAN ARACHNIDA.

By W. J. Rainbow, F.L.S., Entomologist.

The present paper comprises a list of the species of Papuan Arachnida, in the possession of the Trustees. The collection, although numerous in point of specimens, can scarcely be considered *representative* as far as species are concerned. For the most part, the specimens enumerated hereunder have been collected from time to time by Missionaries, few of whom possess that special knowledge so necessary to a successful collector.

Some of the specimens recorded below were collected under the auspices of His Excellency Sir Wm. McGregor, M.D., K.C.M.G., at the St. Joseph's River ; and some were collected by Mr. W. W. Froggatt, at the Fly River, in his capacity as Naturalist to the Geographical Society of Australasia's Expedition in 1885.* The species most common in all collections from New Guinea are those whose arboreal habits or size render them conspicuous, such as *Argyroepeira grata*, Guérin, *A. celebesiana*, Walck., and the huge *Nephila maculata*, Fab.

It is to be regretted that a field so full of interest from a zoological point of view, should be left almost entirely to the enterprise of foreign collectors, and this notwithstanding the fact that it is so close to our doors. Although numerous collectors have at times visited New Guinea, the two principal expeditions from Sydney have been the "Chevert" Expedition in 1875, and the Geographical Society's Expedition, referred to above. The Araneidæ collected by the former were recorded by Mr. H. H. B. Bradley.†

The reader is also referred to a previous paper by me, entitled "Contributions to a Knowledge of the Arachnidan Fauna of British New Guinea."‡

Some of the specimens are vaguely labelled "British New Guinea," whilst others are distinctly located ; thus in the localities recorded below as "British New Guinea, St. Joseph's River," the information intended to be conveyed is—that some specimens are known from the latter locality specifically, and others from British New Guinea generally ; and where the words "British New Guinea" appear between parenthesis, the object is to localise the preceding name.

---

* See Proc. Geogr. Soc. Austr., special vol., 1885.

† Proc. Linn. Soc. N. S. Wales, i., 1876, pt. 2, pp. 137 - 150; pt. 3, pp. 220 - 224, and plates.

‡ *Ibid.*, xxiii., 1898, pt. 3, pp. 328 - 356, pl. vii.

Order ARACHNIDA.

Suborder ARANEÆ THERAPHOSÆ.

Family AVICULARIIDÆ.

Subfamily AVICULARIINÆ.

Genus Phlogius, *E. Simon.*

Phlogius crassipes, L. Koch.

Loc. Fly River (British New Guinea).

Phlogius strenuus, Thorell.

Hab. British New Guinea.

Suborder ARANEÆ VERÆ.

First section ARANEÆ VERÆ CRIBELLATÆ.

Family ULOBORIDÆ.

Subfamily DINOPINÆ.

Genus Menneus, *E. Simon.*

Menneus reticulatus, sp. nov.

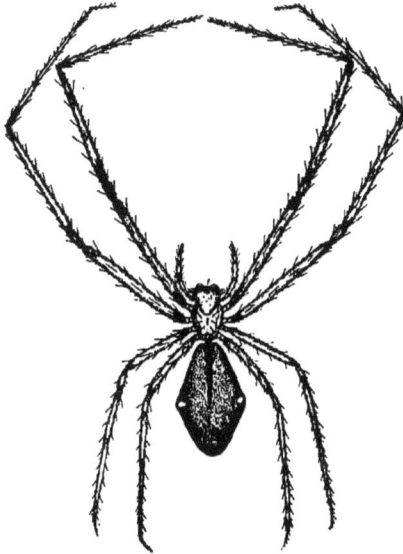

MENNEUS RETICULATUS.

♀. Cephalothorax, 5·9 mm. long, 4·6 mm. broad; abdomen, 12 mm. long, 6·8 mm. broad.

Cephalothorax longer than broad, flat, dull yellowish-brown; sparingly clothed with short yellowish pubescence. *Caput* flat, and level with thoracic segment. *Clypeus* sparingly clothed with yellowish pubescence; broadest in front, retreating gradually to posterior extremity, surface uneven; the median depression is a very short, straight, shallow groove; radial grooves indistinct.

Eyes distributed over three transverse rows of 4, 2, 2; of these the first row is strongly procurved; the lateral eyes of this row are minute, and are seated at the extremity of cylindrical tubercles, springing from the corners of the clypeus; the median eyes are also minute, and are placed closely together, immediately below the two large eyes constituting the second row; those of the third or posterior row are seated well back, and are widely removed from each other.

Legs long, slender, brown, clothed with hoary pubescence, and armed with a few short spines. Relative lengths 1 = 2, 4, 3.

Palpi short, similar in colour and armature to legs.

Falces long, strong, perpendicular, divergent, dark brown, the inner margins fringed with long, yellowish hairs, the outer surface sparingly clothed with greyish pubescence.

Maxillæ dark brown, long, strong, constricted towards the insertion of the palpi, thence greatly divergent, and rounded off at the extremities.

Labium concolorous, rather broader at the base than the apex, which is truncated.

Sternum subtriangular, dark brown laterally, otherwise pale yellowish; the surface pubescent.

Abdomen moderately arched, much longer than broad, angular, narrow in front, but gradually broadening out towards its middle, where it is widest, and thence sharply tapering off to its ultimate extremity; the superior surface and sides are yellow-brown, and reticulated; commencing at its anterior extremity there is a dark brown, rather deep longitudinal median groove of varying width, the margins of which are rough and irregular; the groove terminates at a point a little below the centre; at a distance of rather more than one-third the length of this groove, there are two small, bark brown spots, seated on either side, but quite close to the outer margins; at the widest part of the abdomen there are two lateral obtuse tubercular projections, and these are moderately high; the inferior surface is also yellowish-brown, and reticulated. *Epigyne* a simple, dark brown tubercular eminence, the overhanging lip of which curves slightly upwards towards the centre. *Cribellum* transverse, undivided. *Spinnerets* long and cylindrical.

Obs. An exceedingly interesting species, and the only one obtained by Mr. Froggatt, when out with the Geographical Society's Expedition to New Guinea in 1885. Previously only six species were known, and of these, four were from Australia, one from New Caledonia, and one from Western Tropical Africa. The Australian species are:—*Menneus despiciens*, Camb., *M. unifasciata*, L. Koch, *M. angulata*, and *M. superciliosa*, Thorell, and of these only one, *M. unifasciata* has been found so far south as Sydney. All the Australian species have been described under the generic name of *Avella* (Cambr.).

Loc. Fly River (British New Guinea).

Family PSECHRIDÆ.

Genus Psechrus, *Thorell.*

Psechrus argentatus, Dolesch.

Loc. Fly River (British New Guinea).

Second section ARANEÆ VERÆ ECRIBELLATÆ.

Family THERIDIIDÆ.

Genus Latrodectus, *Walck.*

Latrodectus scelio, Thorell., var. *indica*, E. Simon.

Loc. Fly River and St. Joseph's River (British New Guinea).

Family ARGIOPIDÆ.

Subfamily TETRAGNATHINÆ.

Genus Tetragnatha, *Latr.*

Tetragnatha rubriventris, Dolesch.

Loc. St. Joseph's River (British New Guinea).

Tetragnatha mandibulata, Walck.

Loc. Kiriwina Island (Trobriand), and Louisiade Archipelago.

Genus Argyroepeira, *Emer.*

Argyroepeira celebesiana, Walck.

Hab. British New Guinea, St. Joseph's River.

Argyroepeira grata, Guér.

Hab. British New Guinea, St. Joseph's River.

Subfamily NEPHILINÆ.

Genus Nephila, *Leach.*

Nephila maculata, Fab.

Loc. St. Joseph's River, Kiriwina Island (Trobriand), Engineers' Group (N.E. Coast), and Samarai (Dinner Island), British New Guinea.

Nephila maculata, Fab., var. *pencillum*, Dolesch.

Loc. St. Joseph's River and Fly River (British New Guinea).

Subfamily ARGIOPINÆ.

Genus Argiope, *Aud. et Sav.*

Argiope ætherea, Walck.

Loc. St. Joseph's River (British New Guinea).

Argiope ætherea, Walck., var. *deusta,* Thorell.

Loc. St. Joseph's River, and Kiriwina Island (Trobriand), British New Guinea.

Argiope ætherea, Walck., var. *annulipes,* Thorell.

Loc. Kiriwina Island (Trobriand), and Teste Island (Louisiade Archipelago), British New Guinea.

Argiope picta, L. Koch.

Hab. British New Guinea.

Genus Cyrtophora, *E. Simon.*

Cyrtophora moluccensis, Dolesch.

Hab. British New Guinea.

Cyrtophora viridipes, Dolesch.

Loc. St. Joseph's River (British New Guinea).

Cyrtophora simoni, Rainbow.

Loc. St. Joseph's River (British New Guinea).

Genus Araneus, *Clerck.*

(EPEIRA of Authors.)

Araneus trigonus, L. Koch.

Loc. Fly River (British New Guinea).

Genus Gasteracantha, *Sund.*

Gasteracantha tæniata, Walck.

Loc. Fly River, St. Joseph's River, and Kiriwina Island (Trobriand), British New Guinea.

Gasteracantha crucigera, Bradley.

Hab. Samarai (Dinner Island), British New Guinea.

Gasteracantha hepatica, L. Koch.

Loc. Fly River and St. Joseph's River, British New Guinea.

Gasteracantha pentagona, Walck.

Loc. Fly River, British New Guinea.

Gasteracantha mollusca, L. Koch.

Hab. British New Guinea.

Family THOMISIDÆ.

Subfamily MISUMENINÆ.

Genus Misumina, *Latr.*

Misumina pustulosa, L. Koch.

Hab. British New Guinea.

Misumina innonata, Thorell.

Loc. Fly River (British New Guinea).

Subfamily STEPHANOPSINÆ.

Genus Phrynarachne, *Thorell.*

Phrynarachne tuberculata, sp. nov.

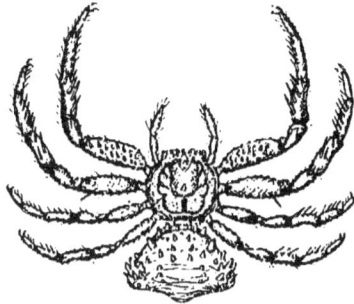

PHRYNARACHNE TUBERCULATA.

♀. Cephalothorax, 4·8 mm. long, 4·8 mm. broad; abdomen, 5·2 mm. long, 7·2 mm. broad.

Cephalothorax almost round, arched, tuberculated, yellow-brown. *Caput* high, arched, uneven, sloping to the front, where it is somewhat obtusely truncated; near the base, which is its highest point, there are two large obtuse tubercles; in front of the latter, and at the sides of the cephalic segment, there are several smaller tubercles; the ocular area, which is of a creamy-white colour, has also a number of small tubercles distributed over its surface, and is moderately clothed with pale yellowish pubescence. *Clypeus* is arched and furnished with a few small tubercles, and the radial grooves are barely distinct; the thoracic segment is yellow-brown, although somewhat darker than the cephalic eminence. *Marginal band* rather broad.

Eyes small, widely separated, but equidistant; these are seated on small tubercles, and arranged in two strongly recurved rows of four each.

Legs yellowish, moderately long, sparingly clothed with yellowish hairs; the tibiæ and metatarsi are each armed with moderately long, but strong yellow spines; all the legs are furnished with tubercles, chiefly of a creamy-white colour, but these are by far the largest and most numerous on the outer surface of the anterior pair; tarsal claws, black. Relative lengths 1 = 2, 3 = 4.

Palpi yellow-brown, aculiate, densely clothed with coarse yellowish hairs.

Falces concolorous, densely clothed with long coarse hairs, cylindriform.

Maxillæ pale yellowish, long, cylindrical, inclining inwards.

Labium concolorous, rather longer than broad, truncated at apex.

Sternum concolorus also, ovate, truncated in front, moderately convex, its surface granulated and sparingly clothed with pale yellowish hairs.

Abdomen sub-pentagonal, truncated in front, slightly projecting over base of cephalothorax, pale yellowish, moderately hairy; the superior surface furnished with tubercles, the largest of which are coniform; in front there is a curved series of eight, and of these the central pair are much the smallest; in an oblique line from the lateral tubercles of the series just described, and running inwards, there is on each side another tubercle, somewhat smaller and more obtuse, and below these again, but in a line directed outwards, there are on each side two smaller somewhat coniform tubercles; commencing at the centre, and running towards the anterior extremity, there is a series of six obtuse tubercles, disposed in three rows of two each; of these, those of the first and second rows are equidistant and small, whilst those of the third or anterior row are somewhat larger and wider apart; at the broadest part of the abdomen, there is on each side a series of seven coniform tubercles, and of these the lateral ones are much the longest; at the sides of and below these again, there are also one or two smaller ones; immediately between the two series referred to above, and close to their inner margins, there are two slightly curved longitudinal rows of three each, the individuals of which are rather small; from the centre, at its widest point, and to its posterior extremity, the abdomen is transversely furrowed, and these furrows are slightly procurved; the inferior surface is of a somewhat lighter colour, and moderately clothed with concolorous hairs. *Epigyne* is a small, black, tubercular eminence.

Loc. Fly River (British New Guinea).

Genus Stephanopsis, *Camb.*

Stephanopsis angulata, sp. nov.

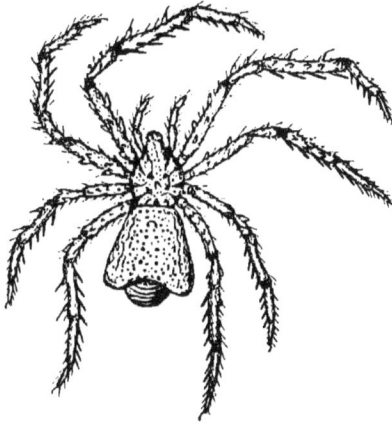

STEPHANOPSIS ANGULATA.

♀. Cephalothorax, 3·9 mm. long, 3·5 mm. broad; abdomen, 4·6 mm. long, 4·2 mm. broad.

Cephalothorax dark brown, nearly as broad as long, arched, granulated. *Caput* granulated, sloping forward, and terminating in an obtuse tubercular eminence. *Clypeus* dark brown, broad, arched, granulated, radial grooves moderately distinct; at the centre there is a small but prominent obtuse tubercle. *Marginal band* broad, granulated.

Eyes arranged in two strongly procurved rows of four each; of those of the anterior row, the median eyes are much the smallest, whilst the lateral eyes, and those constituting the posterior row are of equal size.

Legs long, strong, dark brown, granulated; all are furnished with tubercles, short hairs and spines; but the spines on the underside of the tibiae and metatarsi of the first and second pairs are much the longest and strongest, and are directed forwards. Relative lengths 1, 2, 4, 3.

Palpi short, strong, similar in colour to legs, and densely clothed with coarse hairs.

Falces yellow-brown, cylindrical, clothed with short coarse hairs.

Maxillae dark brown, moderately long, converging inwards.

Labium concolorous, short, broad, narrower at apex than base; apex fringed with coarse hairs.

Sternum yellow, moderately arched, elliptical in outline, and thickly clothed with short yellowish hairs.

Abdomen of a somewhat trapezoidal form, slightly projecting over base of cephalothorax; anterior extremity truncated, posterior extremity strongly indented; the superior surface is of a pale yellowish colour, somewhat uneven, granulated, and slightly arched; in the median line there is a series of six indentations or punctures, and of these the first pair situated at about one-third the distance from the anterior extremity, are the smallest of the series and the closest together; the second pair, situated at about the centre, are much wider apart; those of the third row are about one-third the length from the posterior extremity and rather wider apart than the preceding pair; the posterior angle is somewhat coniform, transversely furrowed and granulated, the furrows recurved; sides rather flat, retreating sharply downwards, pale yellowish, granulated, and furrowed; inferior surface, dull yellowish. *Epigyne* a small, transverse, procurved depression or groove, the inner margins of which are black.

Loc. Fly River (British New Guinea).

Family CLUBIONIDÆ.

Subfamily SPARASSINÆ.

Genus Isopoda, *L. Koch.*

Isopoda herculea, Thorell.

Hab. British New Guinea.

Genus Heterapoda, *Latr.*

Heterapoda venatoria, Linn.

Hab. British New Guinea, Teste Island (Louisiade Archipelago).

Heterapoda venatoria, Linn., var. *foveolata,* Thorell.

Hab. British New Guinea.

Heterapoda salacia, L. Koch.

Loc. Fly River (British New Guinea).

Heterapoda analis, Thorell.

Loc. Fly River (British New Guinea).

Genus Palystes, *L. Koch.*

Palystes pinnotherus, Walck.

Loc. Fly River (British New Guinea).

Family LYCOSIDÆ.

Genus Anoteropsis, *L. Koch.*

Anoteropsis papuana, Thorell.

Loc. Fly River (British New Guinea).

Family OXYOPIDÆ.

Genus Oxyopes.

Oxyopes macilentus, L. Koch.

    Loc. Fly River and St. Joseph's River (British New Guinea).

Oxyopes brevis, Thor.

    Loc. Teste Island, Louisiade Archipelago.

Family ATTIDÆ.

Genus Bavia, E. Simon.

Bavia æriceps, ? E. Simon.

    Loc. St. Joseph's River (British New Guinea).

Genus Icius, E. Simon.

Icius albo-vittatus, L. Koch.

    Loc. Fly River (British New Guinea).

Genus Mævia, C. L. Koch.

Mævia albo-cincinta, Thorell.

    Loc. Fly River (British New Guinea).

Mævia doreyana, Walck.

    Hab. British New Guinea.

Mævia insultans ? Thorell (immature).

    Loc. Fly River (British New Guinea).

Genus Plexippus, C. Koch.

Plexippus dilanians, Thorell (immature, ♂).

    Loc. Fly River (British New Guinea).

Genus Bathippus, Thorell.

Bathippus macrognathus, Thorell.

    Loc. Fly River and St. Joseph's River (British New Guinea).

Bathippus macroprotopus, Pocock.

    Loc. St. Joseph's River (British New Guinea).

Genus Menemerus, E. Simon.

Menemerus paykulli, Aud.

    Loc. St. Joseph's River (British New Guinea).

Genus Ephippus, Thorell.

Ephippus durveilli, Walck.

    Loc. Fly River and St. Joseph's River (British New Guinea).

Order SCORPIONIDÆ.

Family PANDINIDÆ.
Subfamily PANDININI.
Genus Hormurus, *Thorell.*

Hormurus karschii, L. Koch.
Hab. British New Guinea.

Hormurus caudicula, L. Koch.
Loc. St. Joseph's River (British New Guinea).

Order OPILIONIDÆ.

Suborder OPILIONES PLAGIOSTETHI.
Genus Gagrella, *Stol.*

Gagrella hasselti, Thor.
Hab. Fly River (British New Guinea).

A REVIEW OF THE SYSTEMATIC POSITION OF *ZEMIRA*, ADAMS.

By CHARLES HEDLEY, Conchologist.

THE literary history of *Zemira australis* has chiefly consisted of a tossing from genus to genus without reason or explanation. The type species was originally described and figured by G. B. Sowerby, Junr.,[*] as *Eburna australis* from New South Wales. He adds that Dr. Gray considered it to be the *Cancellaria spirata* of Lamarck. This view was upheld by Kiener[†] who reviewing *Eburna*, states that *E. australis* should remain among the *Cancellaria*, where Lamarck had placed it. Deshayes[‡] followed by accusing Sowerby of publishing two names, figures and descriptions for one shell, the first time by drawing it as a *Cancellaria* with three twists on the columella, the second time as an *Eburna* with none. Lamarck and his followers had however been deceived by a

---

[*] Sowerby—The Conchological Illustrations, 1841, Pt. xx., *Eburna*, fig. 5.
[†] Kiener—Coquilles vivantes, n.d. *Eburna*, p. 2.
[‡] Deshayes—Lamarck's Nat. Hist. Anim. s. vert. (2nd ed.) x., 1844, p. 231.

remarkable, though superficial, mimicry, and Sowerby rightly retorted ; "It is impossible that M. Deshayes can have seen the two shells, which are generically and specifically quite distinct."* Reeve supported Sowerby's classification by including *E. australis* in his monograph of *Eburna*,† and added his testimony to the separate existence of the two shells which had confused the Parisian writers. His description but not his figure was copied by Kuster.‡

After thus successfully establishing his species, Sowerby re-described and refigured it as *Pseudoliva australis.*§ The Brothers Adams instituted for *E. australis* a new subgenus *Zemira* which they ranked under *Eburna.*‖ This view is accepted by Tryon¶ but not by Fischer,** who prefers to subordinate *Zemira* to the genus *Macron.* Kobelt, one of the few writers who have contributed more than a copy or a guess to our stock of information, has added to a full account of the shell, a description of the operculum, and concludes that the data presented confirms the classification of Adams.†† Tate has promoted *Zemira* from sub-generic to full generic rank, when describing a second and fossil species.‡‡ The latest classificatory notice is that by Harris§§ who agrees with Tate in considering *Zemira* an independent genus allied to *Eburna.*

No particular argument seems to have been advanced by any-one to show why *Eburna* should be considered the nearest to *Zemira.* The deep canaliculation at the suture, the spotted colour and the general contour certainly present analogies. But except for the plications of the columella, as close a general resemblance is shown by *Cancellaria.* From *Eburna*, *Zemira* differs by its spiral sculpture and especially by the spiral furrow on the fore part of the shell which ends as a projecting point on the aperture.

The dissatisfaction, rather felt than uttered, of authors about the assigned position of the species, is shown by Sowerby's refer-ence of it to *Pseudoliva* and Fisher's to *Macron.*

It has seemed to me that *Zemira* more nearly approximates to the Struthiolariidæ than to the Buccinidæ. The two recent genera *(Struthiolaria* and *Tylospira)* of the former are both ornamented by spiral sculpture ; and in some fossil forms, as

* Sowerby—Thesaurus Conch. iii., 1866, p. 74.
† Reeve—Conch. Icon., v., 1849, *Eburna*, pl. i., sp. 4.
‡ Kuster—Conch. Cab. (2), iii., 1858, p. 84.
§ Sowerby, *op. cit.*, ccxvi., figs. 13, 14.
‖ H. and A. Adams—Gen. Rec. Moll. i., 1853, p. 110.
¶ Tryon—Man. Conch. ii., 1881, pp. 101, 213 ; Struc. and Syst. Conch., ii., 1883, p. 152.
** Fischer—Manuel Conch., 1884, p. 102.
†† Lobbecke and Kobelt—Jahr. deut. Malak. Gesell., 1880, p. 335.
‡‡ Tate—Trans. Roy. Soc. S.A., x., 1888, p. 163.
§§ Harris—Cat. Tert. Moll. Brit. Mus., i., 1897, p. 167.

T. coronata, Tate,* there is a broad and deep channel at the suture. All the members of the Struthiolariidæ have, in the position of the anterior furrow of *Zemira*, some conspicuous mark, either a ridge, a line of tubercles, a depression, or an angle. All have a projection answering to the point on the lip of *Zemira*, which is more or less developed, and attains a maximum in the case of *Struthiolaria calcar*, Hutton.† The feature which I would chiefly emphasise as pointing to the Struthiolariidæ is the broadened and incurved anterior termination of the columella. The southern habitat of *Zemira* agrees better with the distribution of the Struthiolariidæ than with a group so typically northern as Buccinidæ. On the other hand I must admit that though the operculum of *Zemira*, as figured by Kobelt,‡ does not well agree with that of *Eburna*, figured by Adams,§ yet it does not answer to those of *Struthiolaria* figured by Gray,‖ or Smith.¶

Whatever may be the ultimate destination of *Zemira*, there can be no question but that Tate's genus *Eburnopsis*** must accompany it there. According to figures, *Pseudoliva zebrina*, A. Adams,†† bears a marked resemblance to these forms ; but having no personal acquaintance with the species, I forbear to comment further on it.

These notes on the shell characters were put together several years ago. I had hoped that an examination of the animal might prove or disprove the opinion now expressed, but, unfortunately, I have been unable to procure *Zemira australis* in the flesh. From the distribution of dead shells, I conclude that the species lives in depths of a few fathoms on sandy ground. So far as known to me, the range of the species is from Sydney northwards to the Queensland border.

Most authors who have dealt with *Zemira* have coupled it with *Eburna*, a reference as unnatural as that of Lamarck, who called it *Cancellaria*. Fisher's opinion that it is related to *Macron* is more plausible. It is here suggested that to include it in the Struthiolariidæ would harmonise better with the geographical distribution and the shell characters. Information obtainable from the unknown animal may, however, place it in a group of equal value not yet differentiated.

---

* Tate—Trans. Roy. Soc. S.A., xi., 1889, p. 171.
† Hutton—Trans. N.Z. Inst , xviii., 1886, p. 335.
‡ Kobelt—*Op. cit.*, pl. viii., fig. 8.
§ Adams—*Op. cit.*, pl. xi., figs. 5*a*, 5*b*.
‖ Gray—Guide Moll. Brit. Mus., i., 1857, p. 76, fig. 45.
¶ Smith—Phil. Trans., clxviii., pl. ix., fig. 3*a*.
** Tate—*Op. cit.*, p. 117.
†† Sowerby—*Op. cit.*, iii., p. 74, pl. ccxvi., figs. 13, 14.

On Two ADDITIONAL PERFORATING BODIES, BELIEVED
to be THALLOPHYTIC CRYPTOGRAMS, FROM THE
LOWER PALÆOZOIC ROCKS of N. S. WALES.

By R. ETHERIDGE, Junr., Curator.

(Plate xxiii.)

IN 1891 I described* a perforating Thallophyte under the name
of *Palæachlya tortuosa*, occurring in the tissues of a Permo-Carboniferous Monticuliporoid from Queensland, and an Endophytic
form, then believed to be allied to the Saprolegnian Fungi, met
with in the old visceral cavities of another coral *(Stenopora crinita*,
Lonsdale) from rocks of the same age in New South Wales. The
latter was termed *Palæoperone endophytica*.

For the first of these minute and interesting fossils 1 used the
late Prof. P. M. Duncan's genus *Palæachlya*,† proposed by him
for the reception of certain supposed fungal borings detected in
the corallums of Tertiary and Palæozoic corals, particularly
Goniophyllum pyramidale and *Calceola sandalina*.

The recent examination of a well preserved *Favosites*, from the
Devonian Limestones of the Tamworth District, has revealed the
presence of two highly interesting perforating forms, one of which
is, in all probability, allied to *P. tortuosa*, while the other is certainly quite distinct. The second being much the more important,
will be described first.

The tissues of the *Favosites* are penetrated in various directions,
but, more commonly by far, at right angles to the coral's growth, by
longer or shorter chains of moniliform cells (Pl. xxiii., fig. 1), rather
similar to a chain figured‡ by Prof. P. M. Duncan in the tissues of
Goniophyllum. These lines of monillæ divide at irregular distances
apart, either at an acute or obtuse angle, as the case may be, but
no inosculation, contortion, or returning on themselves occur,
although there is a certain amount of curvature. To use an
expression of Prof. P. M. Duncan's, the chains "often dip out of
and come within the focus of the microscope, in their more or less
long course."§ At times they are widely separated, at others
crowded together, the calibre of both the parent portions and
branches being practically the same, the offshoots being quite as

---

* Rec. Geol. Surv. N.S.W., ii., 3, 1891, p. 95.
† Quart. Journ. Geol. Soc., xxxii., 1876, p. 210.
‡ Quart. Journ. Geol. Soc., xxxii., 1876, pl. xvi, fig. 9.
§ Proc. Roy. Soc., xxv., 174, 1876, p. 243.

large as the main portions from which they emanate, but not
again branching. Many of the chains of monillæ appear to have
neither beginning nor ending, but when a termination rises within
the field it is seen under one of two conditions, either as an
ordinary refractive, or an enlarged black globule.

The moniliform cells are usually dark along the margins of the
chain, and refractive in the middle line, but here and there this
refractive centre is absent, when the cells are oblong, and of a
uniform drab-yellow colour. It is only when the cells are destitute
of a refractive centre that they appear to be oblong, whenever
the latter is present they are always strictly moniliform. As a
rule there is no trace of any containing or bounding wall, or of a
sheath, although the monillæ follow one another with great
regularity. Instances do occur, however, in which there appear
to be traces of such a sheath (Pl. xxiii., fig. 4), and in one particular case a chain unquestionably terminates in a clear and
unoccupied tube (Pl. xxiii., fig. 3) ; but this in no way resembles
the tortuous course of *Palæachlya tortuosa*, mihi, or a similar
form to be described later. At intervals of greater or less extent
the continuity of the chain is broken by one, two, three, or more
globules or cells, very much exceeding in size the ordinary monillæ,
and perfectly opaque, in fact quite black (Pl. xxiii., fig. 2). In
only one instance have I observed any deviation from this opacity,
and then the globule was drab-yellow. A chain may either be terminated by one of these black cells ; or, one may be attached at the
side of a chain, out of its alignment, as it were, and similar to a
figure of Duncan's,* who terms it an oospore. One of the chains
without refractive centres is all but terminated with three or more
circular globules united in a cluster (Pl. xxiii., fig. 2), and in the chain
terminating in the clear tube already referred to, there is a similar
cluster, with two single black globules in the course of the chain
also. In a few cases, where the end of a chain has come into view
it merges into an irregular black mass, as seen by Duncan in a
Thamnastræa from the Tasmanian Tertiary.†

On the other hand, no terminal loculus, crowded with zoospores,
as Duncan terms them, and figured by him in *Calceola sandalina*‡
and in *Achlya penetrans*§ has come under notice ; but there
is certainly at one spot a black globule attached to the side
of a chain, from which a rounded mass of pulverulent matter is
proceeding, or is attached. In many of the old visceral chambers
of the *Favosites*, the black globules, Duncan's oospores, may be
seen in a free state, unaccompanied by any moniliform chains.
Another interesting point remains to be noticed—along the edges

---

* Proc. Roy. Soc., xxv., 174, 1876, pl. vii., fig. 48.
† Quart. Journ. Geol. Soc., xxxii., 1876, p. 206.
‡ Quart. Journ. Geol. Soc., xxxii., 1876, pl. xvi., figs. 12 and 13.
§ Proc. Roy. Soc., xxv., 174, 1876, pl. vii., fig. 53.

of the microscopic sections wherever the chains occur, and often well into the sections, among the visceral chambers of the coral, a brown pulverulent substance occurs, always very uniform in colour; does this represent the shed contents of the black globules?

The longest chain observed attained a length of ·5 millimetres, the diameter of the monillæ being ·0075 millimetres, and that of the black globules ·01.

The second form contained within the tissues of the *Favosites* consists of ramifying tortuous tubes, with definite walls, spreading out, returning on their own parts, bifurcating, or forming confused masses (Pl. xxiii., fig. 5). They may be filled with a sherry-yellow, minutely pulverulent matter, or, they may be quite clear of this substance, and only determinable by the presence of the bounding walls, not otherwise differing from the surrounding calcite of the coral, but the outline is very irregular, irrespective of their contorted course. In the majority of instances when these tubes are present, the old visceral chambers of the *Favosites* near at hand are more or less filled with the sherry-yellow pulverulent matter. This material is remarkably like that seen in the perforations of *Palæachlya perforans,** and which Duncan calls tubes "with conidia." In a very few instances I have observed these tubes occupied by patches of dense black matter, similar to the black globular cells of the previously described form. The tortuous nature of this endophyte renders it impossible to speak with any degree of certainty as to the length of an individual tube, but the diameter appears to be tolerably uniform, viz., ·01 millimetres.

I propose to call this organism *Palæachlya torquis*, on account of its much more irregular course. It is otherwise similar in character to, except for smaller dimensions as compared with those of *P. tortuosa*, mihi. *P. tortuosa* is distinctly visible with a one-inch objective (Watson's), whereas the tubes in *P. torquis* cannot be distinguished without the aid of the quarter-inch objective of the same maker. The diameter of the tubes in *P. tortuosa* is ·02 millimetres.

Similar characters separate *P. torquis* from the endophyte figured, but not named, by Waagen and Wentzel,† in the corallites of *Geinitziella columnaris*, Schl. Duncan's illustrations of *Palæachlya perforans* convey, in a general way, the appearance of the tubes in *P. torquis*, allowing for the much more irregular course of them in the latter, and it may legitimately be concluded that, although allied, they are distinct.

As regards the chains of monilliform cells, the probability seems to be that they, and the tubes of *P. torquis*, represent separate

---

* Quart. Journ. Geol. Soc., xxxii., 1876, pl. xvi., fig. 5.
† Pal. Indica, Ser. xiii., Salt Range Fossils, i., 6, 1886, pl. cxv., fig. 1.

C

organisms, notwithstanding Duncan's remark that "whilst recognising two or three forms of parasitic Algæ within these sclerenchymatous structures of recent and ancient date, it does not follow that they are to be made into different species. They may all be parts of the same mycelium-like growth of the parasite, and may depend upon the nature of the nidus in which growth has taken place."[*] In this opinion I am supported by that of my colleague, Mr. Thomas Whitelegge, who has had great experience in the microscopic examination of Cryptogamic life.

Tubes similar to both those now under description have been investigated by many Biologists, with the result of much difference of opinion as to their nature. Prof. John Queckett appears to have been one of the first to investigate similar chains of moniliform cells, and gave an excellent illustration[†] of them permeating the tissues of a coral, at the same time terming them "confervoid growths." He remarked that "confervoid growths also are very frequently met with in the skeletons of corals, as all these bodies possess animal matter, which, decomposing after death, become a nidus for the development of confervæ." In addition to a coral, he figured similar chains permeating the plates of a *Chiton*,[‡] "large canals running through the entire thickness of the sections sometimes preventing the moniliform appearance represented at B" (his fig. 199).

Fuller observations seem to have been made on the simpler tubes, whether of a straight or tortuous nature. Drs. Bowerbank and Carpenter contemporaneously conducted examinations of those permeating the hard parts of Mollusca, and both at first clearly misunderstood their nature. Bowerbank referred to these tubes as "Haversian canals," and speaking of them in the shell of *Ostrea* remarked, "sometimes they pursue their course through this tissue in nearly a straight line for a considerable distance without branching or anastomosing, while in other parts they are tortuous, frequently anastomose, and throw off branches, which have cæcoid terminations."[§]

In the years 1844 and 1847 Dr. W. B. Carpenter examined tubular perforations in the tests of Mollusca. He referred to simple tubes, more or less regularly disposed, and closely resembling those of an ordinary mycelium. In his first Report he said,[||] "The direction and distribution of these tubes are extremely various in different shells; in general they exist in considerable numbers,

---

[*] Quart. Journ. Geol. Soc., xxxii., 1876, p. 209.
[†] Lectures on Histology, ii., 1854, p. 153, fig. 78.
[‡] *Ibid.*, p. 323, fig. 199 B.
[§] Trans. Micro. Soc. i., 1844, p. 139, pl. xvi., fig. 5.
[||] On the Microscopic Structure of Shells.—Brit. Assoc. Report, 1844 (1845), p. 13.

they form a network, which spreads itself out in each layer, nearly parallel to its surface; so that a large part of it comes into focus at the same time, in a section which passes in the plane of the lamina." And again, "I have frequently seen in them indications of a cellular origin, as if they had been formed by the coalescence of a number of cells arranged in a linear direction." These tubes were observed in various Bivalves, particularly *Lima scabrosa*, *Anomia ephippium*, and in species of *Chama*.* Dr. Carpenter's illustrations† convey an excellent idea of some of the tubes in our specimens. Carpenter evidently regarded the tubes as a portion of the Molluscan economy, but later, Kölliker pointed out that all the more or less horizontally spreading tubes described by Carpenter were those of parasites.‡ It is, however, only just to state that Dr. Carpenter was afterward conscious of this, and corrected§ his earlier conclusions.

In 1851 Mr. C. B. Rose investigated‖ tubes perforating the scales of recent and fossil fish, and looked upon them as "infusorial parasites."

Quekett's investigations of shell structure were equally success-ful, for on referring to the subnacreous layer of *Anomia*, *Lima*, and *Arca*, he remarked¶ :—"The tubes sometimes run in a vertical direction, but more frequently horizontally, between or upon the laminæ of which the shell is composed; they are almost always of uniform character, and very frequently branched, so that some of them present very much the appearance of confervæ. . . . Some of these tubes presented a beaded appearance, indicating that they are made up of cells like the tubular fibres of many fungi." Quekett's tubes** are generally similar to those tubes permeating the *Favosites*, but perhaps a little too regular and too much branched, but not so others seen in a Rice-shell.††

In 1858, Mr. C. Wedl described‡‡ tubes traversing the tests of Brachiopoda, Univalves and Bivalves, but his illustrations do not bear particularly on those now under description. He likened them to the living *Saprolegnia ferax*, which he regarded as a Confervan. About the same time Kölliker showed§§ that similar

---

* *Ibid.*, 1847 (1848), p. 100.
† *Ibid.*, 1844 (1845), pl. ix., fig. 20, pl. xviii, fig. 4.
‡ Proc. Roy. Soc. xx., 1859, p. 97; Quart. Journ. Micro. Soc., viii., 1860, pp. 172 and 181.
§ The Microscope, 6th Edit., 1881, p. 382.
‖ *Fide* Duncan, (I have not seen the paper).
¶ Lectures on Histology, ii., 1854, p. 276.
** *Loc. cit.*, fig. 162.
†† *Loc. cit.*, fig. 163 B.
‡‡ Sitz. K. K Akad. Wissensch., xxxiii., 28, 1858, p. 451.
§§ Proc. Roy. Soc., x., 1859, pp. 96 - 98; Quart. Journ. Micro. Soc., viii., 1860, pp. 172 - 181.

tubes existed in the hard parts of Sponges, Foraminifera, Corals, Brachiopoda, and Univalve and Bivalve Shells, and contained sporangia; he regarded them as unicellular fungi. The late Prof. H. N. Moseley appears also to have worked at similar endophytes in 1875, but I regret a want of knowledge of his reference.

The last paper to which I shall refer is Duncan's second communication—"On bodies penetrating Recent and Tertiary Corals," wherein he terms the form *Achlya penetrans*.[*] He remarks that a parasite closely resembling this lived in the tissues of Upper Silurian Corals and Foraminifera, "the main differences between the ancient and modern forms consist in the larger calibre of some of the filaments of the first, their long, often unbranching course, and the frequent development of *Conidia*-looking bodies within them, and the spherical shape of the spores." It does not appear to be quite clear whether Duncan here retains the name he elsewhere proposed for the "ancient" form, viz., *Palæachyla perforaus*, or includes both the "ancient and modern forms" under *Achlya penetrans*.

I am quite in accord with an observation of Mr. A. C. Seward,[†] who says:—"It is generally a very difficult, and often an impossible task, to discriminate between the borings of fungi and algæ in fossil material." In this belief I shall simply leave the tubes described by me as *Palæachlya tortuosa* in the position formerly assigned to them, pending further investigations that future discoveries may afford.

The very much more intricate growth of the tubes described in preceding pages, and their smaller calibre, induce me to consider them as distinct from *P. tortuosa* of the Permo-Carboniferous, and for the sake of clearness they may be known as *P. torquis*.[*]

In considering the systematic position of the moniliform chains of cells, many difficulties present themselves, and in a preliminary investigation of this kind—and it can only be considered preliminary—I merely wish to point out the very strong general resemblance these chains of cells bear to certain unicellular algæ of the group Schizophyceæ, and particularly the Nostocaceæ. The moniliform chains are very like the trichomes of *Nostoc*, allowing for the absence in the former of irregular interlacing, and the enlarged black cells equally resemble the heterocysts of the same genus, which seem to be—so far as my sections enable me to judge—either basal, terminal, or intercalary. Compare the many excellent figures given by Mr. M. C. Cooke, particularly those of

---

[*] Proc. Roy. Soc., xxv, 1876, 174, p. 252.
[†] Seward—Fossil Plants, i., 1898, p. 129.
[‡] *i.e.*, That which is twisted.

Nostoc carneum or *N. commune*,[*] and again *N. hyalinum*.[†] It would, perhaps, be out of place to call attention to the resemblance between endophytic chains of cells of Palæozoic age, with a genus possessing a gelatinous thallus, or envelope, like *Nostoc*, were it not for the fact that Mr. A. C. Seward has collected several instances where the cells of Nostocaceæ in chains have been found in calcareous pebbles at the bottom of lakes in Ireland, and in the State of Michigan.[‡] The genera are *Schizothrix*, *Nostoc*, *Stigonema*, and *Dichothrix*, the first-named enclosed in its comparatively hard tubular sheath. I have already stated that I believed I could distinguish, in more than one instance, a sheath or vagina, enclosing some of the moniliform chains.

In the Nostochineæ, the trichome is either simple or branched; simple in the Nostocaceæ, branched in the Scytonemeæ, etc. In the present instance the trichome is decidedly branched, thus showing a departure from the Nostocaceæ. Furthermore, some genera at least of the Nostochineæ contain marine species.[§] As to the endophytic habit, it is known that species of *Nostoc* occur in the tissues, or mucilage-containing spaces of certain scale mosses.[‖]

In conclusion, and on the whole, it may perhaps be not too much to say that there is evidence of the existence in Palæozoic times of a *Nostoc*-like endophytic alga, which, for systematic purposes, may be known as *Palæopedes*[§] *whiteleggei*. It is named in honour of Mr. Thomas Whitelegge, of the Australian Museum, to whom I am indebted for several valuable suggestions.

---

[*] Cooke—Brit. Fresh Water Algæ, 1882–4, Atlas, pl. xc., fig. 2, pl. xci., fig. 5.

[†] Bennett & Murray—Handb. Crypt. Bot., 1889, p. 131, fig. 359.

[‡] Seward—Fossil Plants, i., 1898, p. 123.

[§] Bennett & Murray—Handb. Crypt. Bot., 1889, p. 441; Seward—*loc. cit.*, p. 122.

[‖] Campbell—Mosses and Ferns, 1895, pp. 115, 117, 119, 122, etc.

[¶] πέδη, a chain or fetter.

ON THE OCCURRENCE OF A STARFISH IN THE UPPER
SILURIAN SERIES OF BOWNING, N. S. WALES.

By R. ETHERIDGE, Junr, Curator.

STARFISH have not so far been recorded from the rich fossiliferous
deposits of Bowning, nor was I cognisant of their presence in
those rocks until Mr. John Mitchell presented a specimen to the
National Collection.

The rarity of this form of life in the Bowning rocks must plead
my excuse for describing so fragmentary an example as that now
referred to. The specimen is interesting, not only on this account,
but also from the fact that it may possibly belong to one of two
by no means common genera of Upper Silurian age—*Palœocoma*,
Salter (*non.* D'Orb), or *Palasterina*, McCoy.

As now preserved, the Starfish consists of portions of three rays
and traces of the interbrachial disk, with the actinial surface exposed. The ambulacra are deep proximally, but become faint
distally. The ambulacral plates are not clearly distinguishable,
but the margins of the valleys are bordered by a row of adambulacral plates, quadrangular and distinct, although the presence of an
outer row is questionable. Combs of rigid spines are attached to
the arm edges, of whatever construction they may be. The mouth
is very large, strongly pentagonal; the oral plates large, triangular,
and apparently of one piece each, instead of two, as should be the
case in a true *Palœocoma*. The arms are united by a disk broken
up by a series of anastomosing lines, giving rise to the appearance
of a polygonal-plated integument when pressed together, but in a
normal condition squamose, as seen through the oral cavity. From
the margin of the disk stream fine long spines that in all probability
covered the whole of the dorsal surface.

It must be at once admitted that, without a more definite
knowledge of the ambulacral plates, and in the face of single instead of double oral plates, the reference of this form to *Palœocoma*,
Salter, is open to doubt ; but the presence of the disk with its
squamose plates, laden with spines, seems to place our fossil nearer
to that genus than to any other. The only other genera known
to me that it appears to approach are *Edrioaster*, Billings ; *Schen-
aster*, M. & W. ; and *Palasterina*, McCoy. As regards the first-
named,[*] the form of the arms, and nature of the disk, are
characters sufficient for separation ; whilst the form of the
adambulacral plates in the second[†] genus are likewise distinct.

* Canadian Org. Remains, Dec. iii., 1858, p. 82.
† Illinois Geol. Survey Report, ii., 1866, p. 277.

With *Palasterina** there is a resemblance in the quadrangular adambulacral ossicles bordering the ambulacra, but the difficulty is increased by the absence of certainty as to the presence or not of a second row of plates, *Plasterina* having but one row, whilst *Palæocoma* possesses two. At the same time there are unquestionably combs of spines along the edges of the ambulacra, which would favour the presence of a second row of plates as in the latter genus. Furthermore, the appearance of the disk is much more akin to that of *Palæocoma* than *Palasterina*, and on the whole it appears to me preferable to refer the Bowning fossil to the former genus rather than to the latter.

A difficulty now presents itself with regard to the name *Palæocoma*. Salter proposed it in 1857,† although D'Orbigny had previously suggested it‡ for the Lias *Ophiura mülleri*, Phillips ; but, according to Zittel,§ even D'Orbigny's name is in part a synonym of *Ophioderma*, M. & T. No other reference to this double use of *Palæocoma*, except that of the late Dr. Thomas Wright,‖ has come under my notice, not even in Dr. B. Stürtz's excellent review of "Fossil and Living Starfish."¶

Under these circumstances, and with the object of avoiding this confusion, I propose to substitute the name *Sturtzaster* for that of *Palæocoma*, Salter, in honour of Dr. B. Sturtz, of Berlin. To the present fossil I propose applying the specific name of *mitchelli*, and if therefore it be correctly referred to *Palæocoma* in the first instance, in the future it must be known as *Sturtzaster mitchelli*.

The specimen is from the Upper Trilobite bed of the Wenlock Series at Bowning, N.S. Wales.

In 1880, the late Prof. Alleyne Nicholson and the Writer proposed** the genus *Tetraster* to take the place of *Palæaster*, Salter (*non*. Hall), Salter's conception of this genus being antagonistic to Hall's later definition.†† More recently Dr. Sturtz has proposed,‡‡ apparently for a similar reason, the name *Salteraster* in the same sense, and to which the date 1886 is attached; it is clear that our name has precedence.

* Brit. Pal. Foss., Fas. i., 1851, p. 59 ; Salter—Ann. Mag. Nat. Hist., (2), xx., 1857, p. 327.
† Salter—Brit. Assoc. Report 1856 (1857), pt. 2, p. 77 ; Ann. Mag. Nat. Hist., (2), xx., 1857, p. 327.
‡ D'Orbigny—Prodrome, 1850, i., p. 240.
§ Zittel—Handb. Pal. i., Abth. 1, p. 445.
‖ Wright—Mon. Brit. Foss. Echinod. Oolitic Form., ii., 1, 1863, p. 29 ; *Ibid.*, ii., 2, 1866, p. 143.
¶ Stürtz—Verhandl. Nat. Vereins Rheinlande, L., 1893, p. 1.
* Mon. Sil. Foss. Girvan in Ayreshire, 1880, pt. 3, p. 324.
† Hall—20th Ann. Report N. York State Cabinet Nat. Hist., 1867, p. 282.
‡ Stürtz—Verhandl. Nat. Vereins Rheinlande, L., 1893, p. 42.

ADDITIONS to the CATALOGUE and BIBLIOGRAPHY OF AUSTRALIAN METEORITES.

By T. COOKSEY, Ph D., B.Sc., Mineralogist.

ADDITIONAL information in regard to the Catalogue and Bibliography of Australian Meteorites, which has came to my knowledge since the publication of the first lists* :—

METEORITES.

BALLINOO. —E. Cohen, Sitz. K. Preus. Akad. Wiss. Berlin, 1898, p. 19.

BEACONSFIELD.—E. Cohen, Sitz. K. Preus. Akad. Wiss. Berlin, 1897, p. 1035, and 1898, p. 306.
Type.—Siderite, belonging to the Octahedrite Group.
Weight.—75 kg.
Loc.—Three kilometres east of Beaconsfield Station, in the Parish of Berwick, County Mornington, Victoria.
Finder and Date.—Mr. Feltus.
Coll.—Dr. Krantz.

CRANBOURNE, No. 2.—
Coll.—" National " should be written for " Technological " in original list.

HAY.—
Type.—Siderolite.
Weight.—9½ lbs.
Loc.—Pevensey Station, Old Man Plain, ten miles below Hay, in a paddock fifteen miles south of the Murrumbidgee River, N.S. Wales.
Finder and Date.—1868 - 1870.
Coll.—In the possession of Mr. Godfrey, of Melbourne.

This Meteorite and that from Eli Elwah† formed, very probably part of the same original mass.

LANGWARRIN.—
Type.—Siderite.
Weight.—17½ cwt.
Loc.—Langwarrin, Victoria.
Finder and Date.—Mr. K. W. Padley, in 1886.
Coll.—Technological Museum, Melbourne.

* Rec. Aus. Mus., iii., 3, 1897, p. 55, and iii., 4, 1898, p. 90.
† See Cat. Australian Meteorites, Rec. Austr. Mus., iii., 3, 1897, p. 57.

MOUNT STIRLING.—A small piece, weighing 14¾ ozs., was found at the same locality and is now in the collection of the Australian Museum, Sydney. It is most probably a portion of the larger mass.

YOUNDEGIN, No. 1.—

Coll.—One piece weighing 22 lbs. is in the Technological Museum, Melbourne, Victoria.

BIBLIOGRAPHY.

COHEN (E.)—Ein neues Meteoreisen von Beaconsfield, Colonie Victoria, Australien.—*Sitz. K. Preus. Akad. Wiss. Berlin*, 1897, p. 1035.

„   Nachtrag zur Beschreibung des Meteoreisens von Beaconsfield.—*Sitz. K. Preus. Akad. Wiss. Berlin*, 1898, p. 306.

„   Ein neues von Meteoreisen Ballinoo am Murchisonfluss, Australien.—*Sitz. K. Preus. Akad. Wiss. Berlin*, 1898, p. 19.

COOKSEY (T.)—The Nocoleche Meteorite, with Catalogue and Bibliography of Australian Meteorites.—*Rec. Austr. Mus.*, iii., 3, 1897, p. 55.

„   Addenda to Catalogue of Australian Meteorites.— *Rec. Austr. Mus.*, iii., 4, 1898, p. 90.

„   Additions to the Catalogue and Bibliography of Australian Meteorites.—*Rec. Austr. Mus.*, iii., 5, 1899, p. 130.

THE QUEENSLAND CATTLE TICK.

By W. J. RAINBOW, F.L.S., Entomologist.

IN the Mémoirs de la Société Zoologique de France, Tome x., 1897, just to hand, Professor G. Neumann publishes a valuable and lengthy paper under the title of " Revision de la Famille des Ixodidés." Amongst the species dealt with there is, of course, that interesting but much dreaded beast, the Queensland Cattle Tick.

According to Neumann, this species is *Rhipicephalus annulatus*, Say, and the synonymy as follows* :—*Ixodes annulatus*, Say ; *Hæmaphysalis rosea*, Koch ; *Ixodes bovis*, Riley ; *I. dugesii*, Mégnin ; *Hæmaphysalis micropla*, Canestrini ; *Boöphilus bovis*, Curtis ; and *Rhipicephalus calcaratus*, Birula. Referring to Say's original description† and the above synonymy, the reviser

---

* Mém. de la Soc. Zool. de France, x., 1897, p. 325.
† Proc. Acad. Nat. Sci. Phil., ii., 1821, p. 75.

appends a foot-note,* of which the following is a free translation :
Say's description is too incomplete to enable one to affirm, with
absolute certainty, that the following forms are synonymous ;
nevertheless it is highly probable that they are.   This is invariably
the case with all the old descriptions of Ixodides.   Say concluded
his description with the following remark :—" Found in considerable numbers on a *Cervus virginianus*, in East Florida."

It is interesting to note that *Rhipicephalus annulatus*, which
is responsible for the transmission of what is known in the United
States as " Texas Fever," is found in Texas, Maryland, Washing-
ton, Chicago, Baltimore, Kentucky, Kansas, Arkansas, New
Mexico, and Honduras, on cattle ; Cuba, cattle and dogs ; Jamaica,
cattle ; Florida, on deer *(Cariacus virginianus*, Bodd.) ; Guada-
loupe, on cattle (where it is known as the " Creole Tick," in
opposition to the *Hyalomma ægyptium*, or " Senegalese Tick ") ;
it is also found in Guatemala, Mexico, and Monte Video ; it
occurs in Paraguay, where it has been found ensconced under
bark of trees, and it has been taken in Timor on the " Sambar "
deer *(Rusa equinus*, Cuvier) ; the cattle of the Caucasus and
Transcaucasus, of Asia, and of Singapore, are also affected by it ;
the pest is also known in North and South Africa ; in Algiers
and at Morocco, on African cattle, Barbary and Touarick sheep ;
again, it occurs in Egypt, Madagascar, at Cape Lopez, Gaboon, and
Port Elizabeth (South Africa).   It may, therefore be considered,
as Neumann observes, cosmopolitan.

The revisor also described a variety from the typical form,
under the name of *Rhipicephalus annulatus caudatus;* but I would
advise those interested in the study of these creatures, to peruse
Professor Neumann's work, from which the notes necessary for
this brief contribution were made.†

It is evident, from the foregoing, that the danger of the
pest spreading is even greater than some of our Australian
authorities and experts may have imagined.   It is only fair,
however, to state that Mr. C. J. Pound, Bacterioligist to
the Queensland Government, has, in a recent paper, drawn
attention to the means by which Cattle Ticks may be spread.
He says‡ :—" Careful and close observations have shown that
although the bovine is the only perfect natural host of the
Cattle Tick, it is only one of the many agencies for its distribution.   It has been proved that the tick will mature, under
favourable conditions, upon the horse and the sheep, and that the
eggs from such ticks are fertile.   I have also found them in
various stages of development attached to goats, kangaroos,
wallabies, and various kinds of birds, as the ibis, crane, peewit,

---

* *Loc. cit.*, p. 407.
† *Loc. cit.*, pp. 324 – 422.
‡ Proc. Roy. Soc. Queens., xiv.. 1899, p. 31.

wild duck, and even on the little shepherd's companion (wagtail)."
In addition to these hosts, Mr. Pound points out that "In studying
the habits of various species of ticks living apart from their host
under natural conditions on some of the northern rivers of this
Colony [Queensland], I have noticed that in the larval stage there
was a natural inclination or instinct to attach themselves to any
moving object, no matter whether animate or inanimate."

The Governments of New South Wales and Queensland have, in
their wisdom, deliminated a boundary beyond which cattle from
affected areas must not pass without inspection and treatment, and
this, so far as it goes, is very right and proper.   It has been urged
that the cattle tick cannot thrive on hosts other than bovines; but
even supposing so, the fact that living examples have, in different
parts of the world, been found upon deer, sheep, dogs, and even under
bark, is in itself sufficiently conclusive evidence as to a means by
which "Texas Fever" may be conveyed, and that, in order to be logical
and thorough, the quarantine regulations should be extended to
all animals travelling from the affected districts.   It is a recognised
fact that the disease is slowly but surely spreading south—hence
the necessity of extending the proscribed area ; and it is only a
question of time, therefore, when it will have invaded New South
Wales, and who can tell where or when its devastating march
will stop ? Stock owners of New South Wales and Victoria would
do well, therefore, to note the facts recorded by Professor
Neumann.

OCCASIONAL NOTES.

I.—*STEGOSTOMA TIGRINUM*, GMEL.

AN ADDITION TO THE FAUNA OF NEW SOUTH WALES.

ON March 14th of the current year, we received from Mr. W.
Hibbs an example of the shark *Stegostoma tigrinum*, Gmel.,*
caught in the River Hawkesbury, New South Wales.   It is a
female, measures four feet in length, and in colour nearly agrees
with var. 3 of Müller and Henle.†   The observations were made
while the shark was still alive, it having been received by us in
that condition.   The ground is creamy, with a greenish hue about
the head and dorsal region ; the markings are black spots, smaller
and regularly arranged on the head, much larger and more widely
spaced on the body and fins.

---

* Gmel.—Linn., p. 1493.
† Müller und Henle—Plagiostomen, p. 24.

The tuberculous ridges are extremely well-marked and are disposed as follows :—The median dorsal ridge commences between the eyes and extends along the edge of the first dorsal fin, thence recommences and similarly passes along the second dorsal ; it once more re-appears and forms the keel of the tail.

On each side of this median ridge and about an inch and a half below it, runs a second ridge which loses itself behind the second dorsal fin, but faintly re-appears on the tail. Another ridge arises above the pectoral, passes along the middle line of the side, and is also traceable along the tail. A fourth ridge commences at the side of the vent and is lost beyond the anal fin. On the median ventral line immediately behind the vent, is another ridge which passes up the edge of the anal fin ; lastly, a ridge leads up to each ventral.

The spots on the tail form regular longitudinal series, one row between each ridge.

The stomach was crowded with a Mollusc, which Mr. C. Hedley recognises as a *Natica*. No trace of the shell was to be seen, but in every case the operculum was present. An examination of the contents of the intestines showed that the operculum is dissolved in its passage, and not ejected from the mouth.

The *Natica* is found on muddy and sandy flats, and the shark passing over such banks must pick up the mollusc by thousands. It evidently crushes the shell, sucks out the animal, and swallows it with the operculum attached. Neither the stomach nor intestines contained any food whatever beyond this particular Gasteropod. Day remarks* :—" The favourite food of this fish is Molluscs and Crustacea."

Although not previously recorded from the Colony, this is the second example known to have been obtained here. On February 14th, 1896, we purchased from a fisherman a specimen caught off Port Jackson.

Hitherto the genus was known in Australian waters only from an example obtained by Mr. Alex. Morton, at Cape York, Queensland. This specimen is also in the Museum collection.

EDGAR R. WAITE.

II.—A SHIPWORM, NEW TO AUSTRALIA.

SOME specimens of "Cobra," received from Captain Almond, Portmaster, Brisbane, prove to be the *Kuphus mannii*, Wright. This species seems not to have been noticed since 1866, when it was described from Singapore. In the same parcel of specimens, which were procured at Cooktown, were included instances of *Calobates thoracites*, Gould.

C. HEDLEY.

---

* Day—Fishes of India, p. 725.

On a FERN *(BLECHNOXYLON TALBRAGARENSE)*, with
SECONDARY WOOD, forming a NEW GENUS, from the
COAL MEASURES of the TALBRAGAR DISTRICT,
NEW SOUTH WALES.

By R. Etheridge, Junr., Curator.

(Plates xxiv. - xxvii.)

The very remarkable and interesting plant remains about to be
described were entrusted to me by Mr. J. Clunies Ross, B.Sc.
(Lond.), of the Technical College, Bathurst, who received them
from Mr. W. Pascoe, the Technological Museum Attendant at
Bathurst. The specimens were obtained from the Coal Measure
strata in the neighbourhood of the Talbragar River, somewhere
between Gulgong and Cockabutta* Hill in the County of Bligh.

The Talbragar or Erskine River rises in the Liverpool Range,
and flowing in a general south-west direction, joins the Macquarie
River a little to the north of Dubbo.

Beds of Permo-Carboniferous age, containing *Vertebraria*, and
probably belonging to the Upper or Newcastle Coal Measures,
have been casually referred to by Messrs. David and Pittman,†
and it is from some portion of these that the fossils about to be
described possibly came.

There are ten specimens, six showing cross or transverse sections
of the stem, with leaves attached, and four in profile, similarly
more or less provided, to say nothing of sundry detached leaves,
in a greater or less state of preservation. I believe these fragmentary remains to be those of a Fern, and shall in consequence
make use of terminology of this section of the Cryptogamia.

In the first specimen the caudex (? or rhizome) is seen in cross
section surrounded by seven radiating fronds, or portions thereof.
(Pl. xxiv. fig. 1.)

The second is a similar fossil, but with eight radiating fronds,
one protruding from below a layer of matrix at a lower level.
The section of the caudex is rather less apparent than in the first
example.

---

* ? Cockaburra, *i.e.*, the "Laughing Jackass."
† Mem. Geol. Surv. N.S.W., Pal. Series, No. 9, 1895, p. ix.

A

In the third specimen there are certainly eight fronds visible, and possibly portion of a ninth, but this example is otherwise particularly valuable for it shows evidence of the minute structure of the caudex; the venation is also remarkably well preserved. (Pl. xxiv., fig. 2.)

The fourth individual is a similar specimen to fig. 2, in that there are the remains of caudex structure, with five or perhaps six radiating fronds, being, in the present instance, impressions of the upper surface (Pl. xxiv., fig. 4). Attention is specially directed to the frond on the upper left hand protruding from below the two fronds immediately above it, and the two on the right hand projecting from a still lower level.

The fifth specimen displays a small caudex surrounded by seven fronds, one of them a young frond, and all again impressions of the upper surface.

The sixth example consists of two small individuals contiguous to one another on the same piece of matrix, one of which is shown in Pl. xxiv., fig. 3. Each possesses three fronds, or portions of three, much shorter and wider than in any of the preceding specimens, and to all intents and purposes pyriform in outline.

The remaining specimens are preserved in profile. The first (Pl. xxv., fig. 5) is a portion of a caudex, with at one end a set of attached fronds, four or perhaps five, forming a kind of corona; and a second series, six or seven in all, at the opposite end, detached and bent backwards out of position, but guided by the evidence of other specimens, there is reason to believe that, although detached, they are practically *in situ*. Immediately above the latter, on one side of the caudex, is a round depression, and half way up it on the other is a small protuberance. Amongst the fronds at the end first described is a small somewhat pyriform scale-like body.

The second specimen seen in profile (Pl. xxvi., fig. 6) is a highly important one, in fact one of the most important of the series, consisting of a short portion of caudex, surmounted by a crown comprising six or seven fronds, and a couple of the scale-like bodies, already noticed in connection with Pl. xxiv., flg. 5. On the face of the crown are what I take to be leaf-scars. On the left of the figure one of the frond petioles is definitely attached to this scar-bearing face, and on the right is a petiole disappearing beneath the matrix, and reappearing beyond in frond form.

The third fossil (Pl. xxiv., fig. 7) is part of a caudex seen partly in a transverse view and partly in profile, in the former case displaying portions of three fronds, radiating therefrom. On the right hand side of the caudex is one of the wart-like protuberances described in Pl. xxv., fig. 5.

The fourth example (Pl. xxvi., fig. 8) is supplementary in some points to Pl. xxv., fig. 5, and is the longest portion of caudex in the collection, but much decorticated. At one end are portions of two fronds extended in opposite directions, and somewhat more than halfway down are traces of two others, one attached, the other protruding through the matrix, and although not attached, as in the first instance, so clearly answering in position to the corresponding frond above, on the right hand side, as to leave little doubt that it also is *in situ*.

Finally Pl. xxiv., fig. 9, is the enlarged caudex of Pl. xxiv., fig. 2, and displays the broken edges of the different zones of the stem which will be explained later on.

The structure of the foregoing specimens may be summarised as follows :—

The Caudex.—The caudex is round, varying from one to three millimetres in diameter, and in length from ten to forty-three millimetres, so far as preserved, sometimes in the round, at other times only as impressions, or both conditions may occur on the same example. When in the former state there is clear evidence of a peeling-off of layers, thus reducing the general bulk of the caudex from what it must have really been in nature. At varied and inequidistant points may be seen the minute thorn-like projections, when a caudex is seen in the round, or, in the case of an impression, as small depressions. I am unable to offer any definite explanation of these, but similar projections have been figured by Mr. R. Kidston on problematical stems from the Lanarkshire Coal-field, called *Psilotites unilateralis*,[*] but as the name implies they are on one side only, nor do I, by calling attention to the resemblance, mean to suggest any relation between the two plants. Mr. A. C. Seward has suggested that these may mark the positions of roots given off from a creeping rhizome.

The mode of distribution of the fronds on the caudex is peculiar, and, so far as the specimens permit me to judge, constant. At intervals occur clusters or tufts of fronds, the intervening caudex surface being destitute of leaf clothing. A caudex is therefore divided into nodes and internodes (Pl. xxv., fig. 5 ; Pl. xxvi., fig. 8). In Pl. xxv., fig. 5, we observe a cluster proceeding from an enlargement or corona, and at the other end a displaced cluster that has been accidentally pressed backwards. In Pl. xxvi., fig. 8, traces of two of these nodes are visible, one a little below the middle of the caudex, the second at the upper end. The enlargement caused by the attachment of the frond bases has almost weathered away, but to the right and left of each the edge of a frond is traceable, particularly on the right of the lower tuft, where it is distinguishable by its revolute margin, and traces of

---

[*] Ann. Mag. Nat. Hist., xvii. (5), 1886, p. 495.

venation. There is a distance of twenty millimetres between the nodes. Pl. xxiv., fig. 7, is on the opposide of the same piece of matrix to that on which Pl. xxvi., fig. 6, is preserved, and the stem of the one is continuous through the shale, and joins that of the other, thereby confirming—first, that the two sets of fronds in Pl. xxv., fig. 5, belong to one and the same caudex, and are looking in opposite directions by accidental displacement; second, that the clusters of fronds, occur at intervals along the caudexes, dividing the latter into nodes and internodes with great regularity.

All the caudexes with fronds attached, as well as sundry small fragments scattered over the matrix of the various specimens, exhibit the remains of internal structure, but in varied degrees of distinctness. From two of the best fragments sections were prepared by Mr. Charles Merton, Section Cutter to the Geological Survey of New South Wales. A general view of one of these, enlarged, is shown in Pl. xxvi., fig. 10, and an enlargement of a portion of the latter in fig. 11 of the same plate. A longitudinal section from another fragment is seen in Pl. xxvii., fig. 12. There is not the slightest shade of a doubt that the portions from which these sections are taken are those of caudexes of the same plant to which the fronds are attached, and not that of any fortuitous intruder. It will be observed that in Pl. xxvi., fig. 10, the centre is occupied by an amorphous mass of opaque material surrounded by a zone of cellular tissue, and two other discontinuous zones. The enlarged illustration indicates that this tissue consisted of radial rows or lines of cells roughly arranged in bundles. Further remarks on these sections will be made later on.

Frond Scars.—The corona, or enlargement, terminating the short caudex impression in Pl. xxvi., fig. 6, bears numerous fronds in various states of preservation. On that portion left bare by the falling off of the latter, are visible triangular frond scars in oblique lines indicating a spiral arrangement of the fronds, precisely as on the caudex of a living Tree-fern. Each of these scars bears a more or less central single pit, indicating the former presence of a vascular opening. On the right-hand side of the figure are broken stipe bases, with portion of a frond protruding through the matrix beyond, whilst on the left-hand are three fronds in a revolute condition, the stipes of two being actually attached to the frond scars. I do not think a more complete demonstration of the relation of these parts, one to the other, could be made.

Fronds.—The fronds are linear-lanceolate, narrow, entire, decreasing very gradually in size towards their apices, which are obtusely pointed. In the young frond (Pl. xxiv., fig. 3) the linear-lanceolate outline gives place to a shorter, broader, and sub-pyriform shape. The fronds appear to have been thick and fleshy ; the longest observed measured twelve millimetres. The proximal end of each is in the form of a broad stipe, articulating to the caudex, and is at once distinguishable from the frond proper by the absence of fascicles, and narrower proportions transversely. The largest number of fronds in any one whorl is twelve, ordinarily there are eight.

The venation is very characteristic and stable throughout the whole of the specimens. A strong midrib, or costa, was present continuous to the apex, but perceptibly decreasing in thickness upwards. The fascicles are free and bilaterally symmetrical, the largest number observed on any one frond being twelve, but the usual number is eight. The veins are short, sub-internal, equal on each side, non-costæform, and rising at a very acute angle. The first bifurcation gives rise to two veinules, which are long and excurrent, following an upward and outward direction, the anterior always the longer of the two. The latter is almost invariably dichotomous, the posterior sometimes so, more often single, the resulting veinlets being short. There is, however, one very characteristic feature—the posterior veinule of the first fascicle on each side is always unbranched, and further, the veins of the first facieles are always the longest in each frond, springing from the costa well within the stripe, and remaining subparallel to the former. The margins of the fronds are at times revolute.

Several microscopic sections of portions of fronds were made with more or less satisfactory results. In Pl. xxvii., fig. 16, which is a section transverse to the line of growth, the general form of the frond is admirably shown, the revolute lateral margins, and the median longitudinal depression occupied by the mid-rib; none of the leaf tissues are preserved. The width of this frond is 1·45 mm., the thickness in the centre 0·17 mm., and the thickness of the ends, including the revolute portions, is 0·3 mm. Pl. xxvii., fig. 15, is a longitudinal section of a frond, or one parallel to its line of growth, and of special interest from the fact that the cellular tissue of the epidermis is to some extent visible, and both the upper and lower surfaces are clothed with setiform hairs. I cannot distinguish either stomata or the parenchymatous mesophyll of the frond. There are certain peculiar and equidistant tissue-pillars, extending transversely for half-way between the upper and lower surfaces, which appear to be composed of much decayed tissue, and enclose clear vacuities that certainly possess determinate margins ; one at the end of the section is filled with amber-brown pulverulent matter. The space below these pillars, extending nearly the whole length of the frond, does not show any regular parenchyma, but has distributed throughout it a number of straight or curved filaments.* A second longitudinal section

---

* These filaments are not unlike the fungal borings described by me under the names of *Palæachlya tortuosa* and *P. torquis* (Rec. Austr. Mus., iii., No. 5, 1896, pp. 121 and 126), but if of this nature, distinct from either of those forms.

(not figured) exhibits the tissue pillars extending completely across the frond from surface to surface, without the intervention of the space just referred to. There are no clear vacuities between the tissue-pillars, but their place is taken by patches of dark brown pulverulent material, as if filling up such hollows. In Pl. xxvii., fig. 14, we see a section taken horizontally through the leaf, immediately below the surface, exposing the mid-rib, veins, and epidermal tissue between the latter, as well as very dark brown round patches between the veins, which occupy the same relative position as the dark spots in Pl. xxvii., fig. 15.

There is evidence of fructification only in the microsections of the fronds, although when the undersides of the latter are visible, and disintegration has taken place, an appearance very similar to fructification presents itself, but that is all. This is due simply to the veinules passing over the revolute margins.

In Pl. xxvii., fig. 16, however, are probably the remains of sori, consisting of a number of filaments clustered under the revolute margins, which remind one of pedicels for the support of sporangia, and attached to one of these on the right-hand side of the frond is a small ovate body that may be a sporangium (Pl. xxvi., fig. 7), very similar to the arrangement of the fructification in *Pteris*.* There is no trace of an indusium. The length of the revolute portion of the frond is 0·3 mm., width of the receptacles containing the pedicels 0·1 mm., its depth 0·06 mm., and the length of the pedicle 0·1 mm.

When first this Fern came under my notice, I took the fronds to be attached in a verticillate manner. I now look upon them as forming small tufts arranged in ordinary close spirals. The structure shown in Pl. xxvi., fig. 6, showing that the fronds were not arranged in a verticil on the same plane, but in a spiral manner, is emphasised by the fact that in Pl. xxiv., figs. 2 and 4, and particularly in the last, some of the fronds appear protruding from below the others. This is specially the case in Pl. xxiv., fig. 4, where the dark shade running across the matrix, indicates a piece removed, displaying a lower level than that to the left of the shading ; on the former are two fronds.

The Scales.—More or less pyriform bodies are visible associated with the fronds in Pl. xxv., fig. 5, and Pl. xxvi., fig. 6 ; these I have tentatively termed "scales." Mr. A. C. Seward, to whom I submitted photographic copies of the present plates, has been good enough to suggest that these may be bulbil-like appendages, or scale-leaves. He remarks that bulbils occur in some recent ferns, such as *Cystopteris bulbifera.* A dimorphic condition of the fronds has been shown to exist in *Glossopteris browniana*,

---

* See Hooker and Baker's Synop. Filicium, 1868, pl. iii., fig. 31.

both by Zeiller and Seward* ; by the latter in examples from the Newcastle or Upper Coal Measures. Those secondary fronds present a scale-like appearance, with an upper convex surface, and slightly spreading and anastomosing veins, but no mid-rib ; the first two characters accord well with the appearance of the "scales" in the present plant. Instances of other recent Ferns possessing two kinds of simple fronds are given by Mr. Seward, in the paper referred to below.

No very satisfactory alliance amongst recent Ferns can be mentioned. All I can do is, as suggested by Mr. Thomas Whitelegge, to call attention to the shrub-like *Oleandra neriiformis*, Cav., in which the fronds are simple-linear-lanceolate, as in our form, subverticillate, and the short stipes articulated with erect frutescent stems.† Except that the fronds here are spiral, and not verticillate at all, there is otherwise a general resemblance between the two. *O. neriiformis* is said by the late Mr. John Smith, formerly of Kew, to be the "only representative of a shrub among Ferns."‡ I believe that some Botanists do not recognise *Oleandra* but merge it in *Aspidium* ; I am, however, content to speak of the plant as referred to by Mr. Smith.

I have been similarly unable to find any near relative of this extraordinary little plant amongst extinct species. The venation is to some extent Pecopteroid, as may be seen by a comparison with the many excellent figures of *Pecopteris* species given by Brongniart in his "Histoire," particularly *P. aquilina*, *P. nervosa*, or *P. cistii*.§

There is a superficial resemblance in the form and venation of the fronds to those of *Marzaria*, Zigno‖ ; but in the latter the frond is pinnate, and the pinnules are described as digito-radiate. Indeed it is the linear-lanceolate form of the pinnules in *Marzaria paroliniana*, and their often radiate arrangement, that first strikes the eye as resembling the fronds of the Australian fossil, especially when the former are pressed from above downwards, in a similar manner to some of those of the latter. The venation of the two forms is almost identical.

Mr. Seward has called my attention to the figures of a Taxodinaceous Conifer, *Cyclopitys nordenskiöldi*, Schml.,¶ from the Russian Permian. In a letter recently received, Mr. Seward remarks :—" In the Russian plant there are apparently no lateral

---

* Quart. Journ. Geol. Soc., liii., 1897, p. 218, pl. xxiii., fig. 1.
† Lowe—Ferns : British and Exotic, 1868, p. 41, pl. xvi. ; Beddome—Ferns of Brit. India, ii., p. 264, pl. cclxiv.
‡ Historia Filicium, 1875, p. 81.
§ Hist. Vég. Foss., i., 1828, pls. xc., xciv., and cvi.
‖ Flora Foss. Form. Oolithicæ, i., p. 168, pl. xix., figs. 3 – 17.
¶ Beiträge zur Jura-Flora Russlands, 1879, pl. xiv., figs. 6 – 8.

veins in the leaves, although it is conceivable that the 'cross-wrinklings' may be veins. I do not think that the two are identical, but the plant is worth referring to. Schmalhausen regards his plant as a Conifer comparable to *Sciadopitys* ('Umbrella Pine')." These are the only comparisons I am able to suggest.

When first dealing with this fossil, I came to the conclusion that it was a Fern of anomalous structure, probably a new genus, but my difficulties were increased on the preparation of the micro-sections of the stem, for I at once saw that the structure revealed was not that of an ordinary Fern. I accordingly forwarded notes and copies of the illustrations to Mr. Seward, who in an exceedingly kind manner has solved my doubts in the letter already referred to, as follows:—"The internal structure strikes me as particularly interesting; your figures 10 and 11 suggest a fairly broad zone of secondary wood—a form of structure practically unknown among recent Ferns, but slightly developed in some species of *Botrychium*, which have undoubted secondary thickening. From Permian and Coal Measure rocks we have, however, several genera of plants which possess characters now shared by Cycads and Ferns, e.g., *Lyginodendron*, *Heterangium*, *Poroxylon*, and others; in the first two the leaves are of the type long known as *Sphenopteris elegans* and other forms, and the stems have a broad zone of secondary wood, with a structure like that of living Cycads. These intermediate types have recently been placed by Potonié in a special class, which he calls Cycadofilices; the genera have been described by Williamson and Scott, Renault, and others. It would seem not improbable that your plant may belong to this class; it certainly suggests a Fern with secondary wood. It would be very interesting to know more about the anatomy, whether the wood consists of radial rows and tracheids separated by *broad* bands of medullary ray tissue—as in Cycads, or, if it is of the more compact form, with narrower and less obvious rays, such as we have in Conifers; also what the tracheids look like in longitudinal section."

Mr. Seward's remarks suggest comparison with *Botrychium*. The structure of the stem in this genus is thus described* by Dr. D. H. Campbell:—"The vascular bundles of the stem are much more prominent than in *Ophioglossum*, and form a hollow cylinder with small gaps only corresponding to the leaves. This cylinder shows the tissues arranged in a manner that more nearly resembles the structure of the stem in Gymnosperms or normal Dicotyledons than anything else. Surrounding the central pith is a ring of woody tissue, with radiating medullary rays, and outside of this a ring of phloem, separated from the xylem by a zone of cambium,

---

* Mosses and Ferns, 1895, p. 243.

so that here alone among Ferns the bundles are capable of secondary thickening. The whole cylinder is enclosed by a bundle-sheath
(endodermis) consisting of a single layer of cells. The cortical
part of the stem is mainly composed of starch-bearing parenchyma,
but the outermost layers show a formation of cork." An excellent
diagramatic sketch of the several parts accompanies these remarks.

I believe Pl. xxvi., fig. 10, to practically represent the greater
portion, if not all, of the stem or caudex, viewed transversely. It
will be noticed that the central portions retain a fairly continuous
oval contour, but the outer portions, possibly from extraneous
causes, have been crushed together, and the contour broken or
distorted. The dark centre in our figure, and from which the
whole structure in the specimen has been obliterated, represents
without doubt the pith (there is no evidence of primary wood),
whilst the zone surrounding this is the secondary wood or xylem.
The two outer rings in Pl. xxvi., fig. 10, judging by Dr. Campbell's,
may possibly represent—the inner one the endodermis, and the
outer the cork formation of the parenchyma. These rings in the
fossil are of a dark orange-brown colour. In the enlarged figure,
(Pl. xxvi., fig. 11) the dark radiating lines perhaps represent the
medullary rays.

The following measurements were kindly made by Mr. T.
Whitelegge :—

Longer diameter of caudex	1·8 mm.	
Shorter „ „	1·2 „	
Diameter of pith	0·35 „	
Width of ring of secondary wood	0·25 „	
Space between exo- and endoderm... ...	0·1 to 0·2 mm.	
Space between endoderm and secondary wood...	0·0 to 0·1 „	
Width of exoderm	0·04 mm.	
Width of endoderm...	0·03 „	

In a longitudinal section of the caudex (Pl. xxvii., fig. 12) the
same number of zones can be distinguished as in a transverse
section (Pl. xxvi., fig. 10). Thus, the central cylinder, without
structure, is followed by the zone of secondary wood, in which
faint longitudinal parallel lines can be discerned, answering to the
radial lines in Plate xxvi., figs. 10 and 11, but no minute details
can be made out. The edge of the secondary wood is of the same
deep amber-brown colour already referred to in other parts of the
organism, and between this and the layer corresponding to the
endoderm in Dr. Campbell's diagram of the stem in *Botrychium*,
is a further narrow structureless zone (the inner of the two in
Pl. xxvi., fig. 10), that varies so much in radial diameter. The
endoderm and the exoderm are again of a deep amber-brown tint,
and form strongly marked features of the section, the intermediate

space between them being again structureless. This section, which was prepared from a stem fragment enveloped in matrix in the hope that it would display the longitudinal structure simply, has by accident revealed other unexpected details, in the form of three bract-like bodies on each side, opposite to one another, two and two, those on the right being better preserved than those on the left. As sectioned they form an extension of the parenchymatous zone, and are margined by a continuation of the amber-brown exodermic layer, which is cellular, the tissue having the appearance of the epidermal of the fronds. The two lowest bodies are the longest and best preserved. The only suggestion I can offer of this structure is that the section at this point traverses a node, and that we here see petioles of some of the fronds.

In the same section, but detached from this caudex fragment, is what may well be termed a root and rootlets (Pl. xxiv., fig. 13). The former is straight, 2·7 mm. long, with a transverse measurement of 1·5 mm., and from it on each side are given off at right angles longer and shorter processes, varying in length from 0·15 to 0·3 mm. There is no structure preserved. Whether or no these are a root and rootlets of the plant under consideration, it is of course impossible to say.

Could further points of structure be made out in this interesting fossil, a comparison might then be instituted with that of *Lyginodendron*, Will., on the one hand, and that of *Heterangium*, Corda, on the other. On a superficial comparison with figures of both,* a general resemblance is noticable, particularly in the central cylinder, and the surrounding zone of secondary wood, but as we are unacquainted with the constitution of the central cylinder, whether of pith and primary wood, as in *Lyginodendron*, or primary wood alone, as in *Heterangium*,† it is impossible to carry the comparison further.

One very interesting point, however, remains to be referred to— the attachment of a fern foliage to stems with affinities of a higher order. The late Prof. W. C. Williamson suggested ‡ that the rachises of certain ferns known as *Rachiopteris aspera*, from a similarity in some of their tissues to those of *Lyginodendron*, were the petioles of the leaves of that genus. He remarked—" If we are correct in this supposition, we have now, for the first time, in *Lyginodendron Oldhamium*, a Fern of which the stem or rachis exhibits a highly developed form of exogenous growth. . . Some months ago Mr. Kidston sent me some stems which he believed

---

* Williamson—Phil. Trans. for 1873, clxiii., pl. xxii., fig. 1 ; Williamson and Scott—*Ibid.* (B) for 1895, clxxxvi., pl. xviii., fig. 1; Seward—Ann. Bot., xi., pl. v., fig. 1.
† Williamson and Scott—*Loc. cit.*, clxxxvi., p. 745.
‡ Phil. Trans. (B) for 1887, clxxviii., p. 298.

to belong to *Sphenopteris elegans*, the cortex of which displayed an exactly similar series of thickened horizontal parallel bands. Still more recently, he received from my friend Professor Von Weiss, of Berlin, and forwarded to me, a beautiful specimen of an exactly identical stem, attached to which are the unquestionable pinnules of *Sphenopteris elegans*. As far as these internally structureless specimens affect the question, they suggest the possibility that both the species of *Heterangium* may also prove to be Ferns." Again, speaking of the two genera already referred to, Prof. Williamson remarked*—"One thing is certain, viz., that in their internal organisation they present combinations of tissues that find no representatives amongst living plants. Possibly they are the generalised ancestors of both Ferns and Cycads, which transmitted their external contours to the former, and their exogenous modes of growth to the latter types. In considering this possibility, we must not forget that in *Strangeria* we have a still living plant in which the stem of a Cycad bears fronds, the leaflets of which retain the dichotomous nervation of a true Fern. The *Strangeria* has retained, not only the primitive exogenous stem of some ancestral type, in common with its other Cycadean relatives, but also the peculiar Fern-like leaflets, which may also have come down to it from Palæozoic times. Hence we have here a combination of Fern-like features and of an exogenous mode of growth. Such being the case, it need not startle us if we have to conclude that a similar cambination existed during the Carboniferous age." On this subject Messrs. Williamson and Scott remark conjointly†—"In all cases where the petioles can be determined as belonging to *Rachopteris aspera*, we now know that we have to do with the foliage of *Lyginodendron*," thus confirming previous conclusions, "namely, that the leaf would fall under the form-genus *Sphenopteris* of Brongniart, as shown by the finely cut foliage and the acute angles between the veins. . . The mere fact that the foliage of *Lyginodendron* resembled that of certain Ferns is in itself no proof of affinity with *Filices*. The classical case of *Strangaria* is a sufficient warning against any such hasty inference. It must, however, be remembered that in the foliage of *Lyginodendron* we have not only fern-like *form* and venation, but also fern-like *structure*, whereas in the case of *Strangeria*, a single transverse section of the petiole would be sufficient to prove that the plant is no Fern but a Cycad."

The form of the leaf in the present fossil is certainly that of a fern, but unfortunately the structure is not in a sufficiently good state of preservation to warrant any definite generalisations. There is certainly no evidence of the existence of palisade parenchyma; on the other hand the presence of a bifacial structure

---

* Phil. Trans. (B) for 1887, clxxviii., p. 299.

† Phil. Trans. (B) for 1895, clxxxvi., p. 727.

and epidermis seems to be tolerably apparent, and the fact that the fronds were supplied by a single vascular bundle, as in *Heterangium*.[*]

As regards the petioles a general resemblance exists between those attached to our longitudinal section Pl. xxvii., fig. 12, and that of *Lyginodendron* given by Williamson and Scott,[†] but little or no minute structure can be made out. They are here opposite, and not spiral as in Pl. xxvi., fig. 6; they are spiral in *Heterangium*.[‡]

In conclusion, as to the general affinities of this very interesting little plant the following observations may not be inappropriate. Messrs. Williamson and Scott remark[§]—"The occurrence of secondary thickening in a Fern-like plant is not in itself very surprising. We know that it takes place in a perfectly typical way, though not to any great extent in the stems of *Botrychium* and *Helminthostachys* at the present day." The same may be justly claimed for the present plant.

It is unnecessary to follow Messrs. Williamson and Scott through their very interesting line of reasoning to show the structural connection of *Lyginodendron* and *Heterangium*, with both Ferns and Cycads, but the following sentence is probably very pertinent to the Talbragar fossil—"The view of the affinities of *Lyginodendron* and *Heterangium*, which we desire to suggest, is, that they are derivatives of an ancient and 'generalised' (or rather non-specialised Fern-stock), which already show a marked divergence in the Cycadean direction," and they think " the existence of a fossil group on the borderland of Ferns and Cycads is now well established."[¶] For this intermediate group of plants Dr. H. Potonié has proposed [**] the divisional name of Cycado-filices, a class not hitherto recognised, Mr. Seward remarks to me, in the Southern Hemisphere.

I intended using the generic name of *Pteroxylon* for this plant, but Mr. J. H. Maiden, Director of the Botanical Gardens, Sydney, informs me that as *Ptaeroxylon* it was employed in 1835 for the "Sneezewood" of South Africa. As, however, it is very desirable to retain in the name a connection between the presence of secondary wood and a Fern alliance I have adopted a suggestion made to me by Mr. Whitelegge, and term it *Blechnoxylon*.[††] Now, although βλῆχνον, is literally a "kind of fern," still, according to Loudon it is also "one of the Greek names of the fern,"[‡‡] and may

---

[*] Phil. Trans. (B) for 1895, clxxxvi., p. 734.
[†] *Ibid.*, pl. 26, fig. 22.   [‡] *Ibid.*, p. 756.
[§] Phil. Trans. (B) for 1895, clxxxvi., p. 766.
[¶] *Ibid.*, p. 769.   [¶] *Ibid.*, p. 770.
[**] Lehrbuch der Pflanzenpalaeontologie, Hief 2, 1897, p. 160.
[††] βλῆχνον and ξύλον.
[‡‡] Encyclopædia of Plants, 1880, p. 881, Note 2183.

in consequence, I think, be justifiably used in the sense intended. The plant will therefore in future be known as *Blechnoxylon talbragarense.*

The fossils are associated in the same deposit with leaves of *Glossopteris*, and stems of our characteristic Coal Measure Conifer, *Brachyphyllum.*

Throughout this enquiry I have been very ably assisted by my Colleague, Mr. T. Whitelegge, and desire to take this opportunity of expressing my thanks not only to him, but also both to Mr. E. R. Waite, who has spared no pains to render the illustrations accurate and intelligible, and to Mr. J. P. Hill, B.Sc., of the Biological Laboratory, Sydney University, for the loan of micro-preparations of *Blechnum, Strangeria,* and other plants.

DESCRIPTIONS OF TWO BEETLES FROM MOUNT KOSCIUSKO.

By W. J. Rainbow, F.L.S., Entomologist.

In working over the collection of Australian Carabidæ contained in the cabinets of the Australian Museum, I came across two species apparently undetermined—one a *Percosoma*, and the other *Notonomus*, sp. These are, therefore, now described.

Some time ago, Mr. T. G. Sloane described the Australian and Tasmanian forms of the genus *Percosoma* as known to him.[*] Of these *P. montanum*, Casteln., and *P. concolor*, Sloane, were recorded from Victoria ; the former from Yarragon, Gippsland (Sloane), Dandenong Ranges (French), and the latter from Marysville District (Track to Yarra Falls, Best). Two others, *P. carenoides*, White, and *P. sulcipenne*, Bates, were from Tasmania. The four species here enumerated comprised all that was known of the Australian *Percosoma* up to the date of the publication of Mr. Sloane's paper, and from then until now, no further additions to our knowledge of the native species of this genus have been made.

In the working out of the species *(Percosoma)* herein described, I have been courteously assisted by Mr. Geo. Masters, Curator of

---

[*] Proc. Linn. Soc. N.S.W., vii., 1892, pp. 60 - 62.

the Macleay Museum, and I therefore have much pleasure in associating his name with it, specifically.

Genus Percosoma, *Schaum.*

PERCOSOMA MASTERSI, *sp. nov.*

(Fig. 1).

Measurements:—Head (to tips of mandibles) 5 mm. long, 3·5 mm. wide; pro-thorax, 6 mm. long, 6·9 mm. wide; elytra 11·7 mm. long, 7·3 mm. wide.

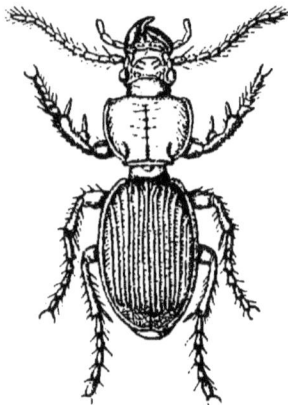

Fig. 1.

Black, shining, narrow, elongate. *Head* rather large, smooth; transverse impression behind the eyes strongest laterally, faint behind vertex; vertex rather flat, frontal impressions well defined, short, anterior angle with a row of six punctures; jaws not long, hooked at apex; eyes prominent, inclosed behind, projecting before post-ocular prominences—these not strong. *Antennæ* subfiliform, clothed with pale, yellowish bristles; basal joint has a deep round puncture above. *Prothorax* slightly convex, cordate, truncate in front and behind, median line lightly marked; sides slightly rounded on anterior half, gently narrowed but not sinuate towards base; lateral border narrow, extending from anterior to posterior angles—the latter somewhat obtuse. *Elytra* convex, oval, narrow, striate (the striæ—seven upon each elytron—very distinct), intermediate spaces flat; shoulders rounded; base declivous to peduncle, apex broadly rounded; lateral border narrow; within the outer margin of each elytron, there is a row of small punctures, widely separated from each other individually; of these four are seated well forward, one at the centre, and the remainder (four) towards the apex, the last three rather closer together. *Ventral surface* lightly rugose laterally. *Anterior legs*—thighs canaliculate below, dilatate at middle; outer angle of tibia finely serrated.

Obs.—Judging from the fact that the anterior thighs are similar to those of the female of *P. montana* as described by Sloane;* the species diagnosed above is doubtless a female also.

---

* *Loc. cit.*, p.₁60.

Genus Notonomus, *Chaud.*

NOTONOMUS MONTANUS, *sp. nov.*

(Fig. 2).

Measurements :—Head (to tips of mandibles) 5·5 mm. long, 4·2 mm. broad ; pro-thorax 5·1 mm. long, 5·8 mm. broad ; elytra 14 mm. long, 7·5 mm. broad.

Black, shining. *Head* smooth, broad, frontal impressions well marked, clypeal suture distinct, ending on each side in the frontal impression ; eyes somewhat prominent, inclosed behind. *Pro-thorax* broader than long, grooved down the middle, slightly rounded on the sides, gently narrowing towards the base ; posterior angles rounded off, lateral impressions short ; marginal border reflexed on the sides, reaching as far but not extending beyond the inner side of the lateral impressions on each side of the base ; posterior marginal punctures in the lateral border at the basal angles. *Elytra* oval, not convex, a little narrower towards the base, rounded on the sides, broadest just beyond the middle, sinuate behind, dehiscent at apex, sides and apex declivous, striate ; interstices flat, the ninth marked throughout its course with umbilicate punctures, those towards the base and apex close, but not confluent ; lateral margins wide, humeral angles not marked. *Abdominal segments* normal.

Fig. 2.

Obs.—The third interstice of each elytron has a series of punctures, but as the number is not uniform, there being six on the right elytron and four on the left, these impressions can scarcely be accepted as specific characters.

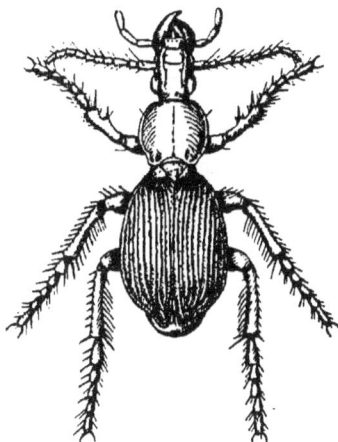

LARVA and PUPA of *BATOCERA WALLACEI*, THOMS.

By W. J. Rainbow, F.L.S., Entomologist.

A SHORT time ago, the Trustees of the Australian Museum were presented with a small collection of Insects from Samarai, or Dinner Island, British New Guinea. These had been gathered from time to time by the donor, the Rev. C. W. Abel, in the field of his labours, and amongst them were included larval and pupal forms of that huge Longicorn, *Batocera wallacei*, Thoms.

In respect of the former, which is apparently nearly fully grown a brief description may be interesting. The animal is nearly four inches long, footless, gradually tapering posteriorly, as is most frequently the case with Longicorn larvæ ; the head is short, broad, flat, corneous, punctate, black, and provided with short, strong, incurved mandibles ; the pro-thoracic segment is much larger than those succeeding, and of a glossy mahogany-brown hue, the anterior, posterior and lateral angles, both above and below, are dirty yellowish-brown, and rough ; the other segments are fleshy, yellowish, and all, with the exception of the two last, provided both above and below, with large, rough, transversely oval, granulated patches ; these are, of course, the organs of locomotion.

The pupa is large, soft, and of a greenish hue, with yellowish dorsal and lateral patches ; in other respects it presents the usual appearance of Longicorns at this stage of their existence. The wings and legs are folded against the sides, and the feet doubled under ; the enormous antennæ are turned back against the sides of the body, the third and succeeding joints being coiled round and round, something like the mainspring of a watch.

The adult insect is so well known, that there is no need to to describe* it here.

---

* See Arch. Ent., i., p. 447, pl. xviii., fig. 1.

DESCRIPTIONS OF NEW LAND SHELLS, WITH NOTES ON KNOWN SPECIES.

By CHARLES HEDLEY, Conchologist.

(Plate xxviii.)

PAPUINA MAYANA, *sp. nov.*

(Pl. xxviii., figs. 10, 11).

Shell imperforate, ovately conical, periphery rounded, glossy. *Colour*—the base and a subsutural stripe in the lower three whorls are ochre-yellow, contrasting sharply with a broad dark chocolate band which intervenes, the upper whorls are slate. *Whorls* six, rounded, divided by an impressed suture. *Sculpture* oblique, regular, incremental lines are decussated by faint, spiral striations the latter only visible under the lens. *Aperture* very oblique, slightly descending, subrhombic ; lip a little reflected ; columella deeply entering, then straight, edged within, not truncate anteriorly but joining the basal lip at an angle ; a thin white callus spreads on the base. Major diameter 22, minor 19 mm.; height 25 mm.; another specimen, 18, 22, 23·5 mm.

The species has a general superficial resemblance to *P. meta* from the Solomons. The Australian *Papuina* are confined to the Torresian Region, of which they are characteristic inhabitants. At present there are known, *P. macgillivrayi*, Forbes ; *P. bidwilli*, Pfeiffer ; *P. ccrea*, Hedley ; *P. poiretiana*, Pfeiffer ; *P. fucata*, Pfeiffer; *P. conscendens*, Cox; and *P. folicola*, Hedley, The novelty is a near ally of *P. poiretiana*, from which it differs by colour, greater breadth, and absence of perforation.

Loc.—Collected by Miss E. Hatfield at Rossville, on the Upper Annam River, near Cooktown, Queensland.

It is named in honour of Dr. T. H. May, of Bundaberg, at the desire of Mr. Arthur Dean who presented the type specimens to the Trustees.

ENDODONTA ACULEATA, *sp. nov.*

(Pl. xxviii., figs. 1, 2, 3).

Shell thin, depressed, spire level, umbilicus a quarter of the shell's diameter. *Colour* pale ochraceous. *Whorls* three and a half, rounded. *Sculpture* oblique, thin, recurved, epidermal lamellæ, in number about thirty, cross the last whorl from the suture to the umbilicus. Each lamella is produced at intervals into long, slender points, so arranged as to fall into four equidistant spiral lines, one being on the base, one at the periphery,

B

and two above.   On the inner whorls the lamellæ grow less prominent and disappear.   Between the lamellæ the surface of the shell is microscopically spirally grooved.   *Aperture* oblique, not descending, unfinished.   A slight callus spreads in advance of the aperture on the previous whorl.   Major diameter, 2·4, minor 2·2; height 1·2 mm.

The curiously tagged lamellæ sufficiently distinguish this species from all co-generic forms.   In systematic order its place is next to *E. paradoxa*, Cox.

Loc.—Two specimens were collected at Wollongong, New South Wales, by Mr. A. E. Lower.   That figured has been presented by Dr. J. C. Cox.

ENDODONTA NORFOLKENSIS, *sp. nov.*

(Pl. xxviii., figs. 4, 5, 6).

Shell minute, depressed globose, perforate.   *Colour* uniform tawny-olive.   *Whorls* five, gradually increasing, rounded, separated by a deep suture.   *Sculpture*—fine close longitudinal riblets cross the whole breadth of each whorl.   *Aperture* perpendicular, lunate, outer lip sharp.   *Armature*—two lamellæ on the parietal wall, one median, curling within and projecting beyond the aperture, the other straight, posterior and deeper seated; on the outer wall two lamellæ are so situated as to divide with the parietals the height of the aperture into fourths; on the inner side of the columella is a short curved ridge ; the armature does not extend back into the shell for more than half a whorl, its rear aspect is shown in the accompanying dissection (fig. 5).   Base rounded, umbilicus about a fifth of the shell's diameter, exposing the previous volutions.   Major diameter 1·44, minor 1·28 ; height 1 mm.

The mouth armature will readily distinguish this species, which belongs to the subgenus *Thaumatodon*, Pilsbry, and is one of many links between the faunas of Norfolk Island and New Zealand.   Allusion was made to this species by Brazier in describing *E. dispar* from Tasmania.*

The land shells known from Norfolk Island have been described by authors as follows:—*Helix phillipii*, Gray; *H. campbellii*, Gray; *H. insculpta*, Pfeiffer; *H. flosculus*, Cox; *H. quintalæ*, Cox; *H. exagitans*, Cox; *Carocalla stoddartii*, Gray; *Omphalotropis cerea*, Pfeiffer; *O. albocarinata*, Mousson; *Palaina coxi*, H. Adams; *Helicina pictella*, Pfeiffer; *H. norfolkensis*, Pfeiffer.   Brazier has also recorded the introduction of the European *Vallonia pulchella*.†

Loc.—Collected by J. Brazier in 1865 on Norfolk Island, South Pacific, in a guava forest on a hill side.

---

* Brazier—Proc. Zool. Soc., 1870, p. 661.
† Brazier—Journ. Conch., 1879, p. 281.

DENDROTROCHUS MENTUM, *sp. nov.*

(Pl. xxviii., figs. 12, 13).

By colour, texture and general shape this species might be taken for a diminutive form of the well known *D. helicinoides*, Hombron and Jacquinot, of the Solomon Islands. The chief distinction between the two, lies in a thickened, curved ridge, analogous to that of certain species of *Papuina*, situated behind the aperture of *D. mentum*. This ridge is a constant feature in a fair series before me, supported by the difference in size, *D. mentum* being three quarters of that of *D. helicinoides*, and the difference of locality, sufficiently isolates the novelty to require a name for it. Major diameter 12, minor 10 ; height 9 mm.

D. pyxis, Hinds, from the adjoining island of New Ireland, also is contracted and then inflated behind the aperture, though not so sharply so as *D. mentum*. In the species now recorded, the genus has its most western outlier. It was demonstrated in a previous article in this serial,* that *Dendrotrochus* was related not to *Papuina* but to *Trochomorpha*. The group seems worthy of generic recognition.

Loc.—Ralum, New Britain ; presented by Madame E. E. Kolbe.

TORNATELLINA WAKEFIELDÆ, *Cox.*

(Pl. xxviii., fig. 14).

Achatinella wakefieldæ, Cox, Mon. Australian Land Shells, 1868, p. 78.

Specimens of this unfigured species having recently been received by the Trustees from Bryon Bay, New South Wales, the opportunity is taken to illustrate it.

In reviewing the above quoted work, G. W. Tryon pointed out that the species could not be included in *Achatinella.*† Crosse followed by suggesting that it would be more appropriately placed in *Tornatellina*,‡ an opinion which was echoed by Brazier.§

Petterd reported the species as occurring in rotten wood near Lismore and on the Clarence River, New South Wales.|| H. Tryon found it on orange trees in the vicinity of Brisbane, Queensland.¶

---

* Hedley—Rec. Austr. Mus., ii., 1895, p. 90.
† G. W. Tryon—Am. Journ. Conch., iv., 1868, p. 285.
‡ Crosse—Journ. de Conch., xvii., 1869, p. 176.
§ Brazier—Proc. Zool. Soc., 1872, p. 807.
|| Petterd—Journ. de Conch., xxv., 1877, p. 361.
¶ H. Tryon—Report on Insect and Fungus Pests, No. 1, 1889, p. 148.

PAPUINA IIINDEI, *Cox.*

(Pl. xxviii., figs. 7, 8, 9.)

Cochlostyla Hindei, Cox, Proc. Linn. Soc. N. S.W. (2), ii., 1888, p. 1063, pl. xxi., figs. 1, 2.

Helicostyla hindei, Pilsbry, Man. Conch., ix., 1894, p. 229.

The most eastern point to which on assured, that is anatomical, grounds, the tribe of Belogona has been traced, is Waighiou, Dutch New Guinea, in the form of *Helicostyla conformis*, Férussac. On other and doubtful grounds this tribe is assumed to occur as far east as the Solomon Islands, in *Crystallopsis*, Ancey. With a view to deciding on more exact limits, I wrote to Madame Kolbe, of Ralum, New Britain, enclosing a coloured drawing of *Cochlostyla hindei*, and requested her to procure the animal for study. Madame Kolbe, always a generous friend to this Institution, kindly responded by forwarding some well preserved material. The result of an examination of this is as follows :—

Jaw (fig. 7) thin, ends rounded, smooth, centre crossed by weak ribs.

Radula (fig. 9) of the type which Pilsbry has described for *P. vexillaris*, with rows meeting in the centre at an acute angle, and with enlarged and broadened cusps.

Genitalia (fig. 8) without accessory organs on vagina. Tentacle retracted between the branches. Penis-sac long, flattened, without internal papilla. Retractor muscle attached to floor of lung cavity. Epiphallus long, slender and twisted. Spermatophore on long slender duct.

This dissection must certainly effect the transference of *Cochlostyla hindei*, from *Helicostyla* to *Papuina* as those genera are arranged in Pilsbry's classic "Guide to the Study of Helices." On this point the absence of a dart sac is conclusive and is supported generally by the structure of the male organ and of the dentition. Credit must be given to Brancsik, who in describing *H. heimburgi*, synonymous as Pilsbry suggests with the species before us, correctly indicated its systematic position by referring it to *Geotrochus*. I regard *P. hindei* as related to *P. xanthochila*, Pfr.

NOTE ON *SCYLLARUS SCULPTUS*, LATREILLE.

BY THOMAS WHITELEGGE, Zoologist.

(Plate xxix.)

THE history of this beautiful, rare, and well-marked species of *Scyllarus* may be briefly given, as follows :—

It was figured by Latreille* with a bare name, which occurs on page 5, in the explanation of the plate of the undermentioned work. At a later date† the species is again referred to, the reference consisting of a line or two, stating that the figure on P. cccxx., represents *Scyllarus sculptus*. The next notice, as far as I can ascertain, is by M. Milne Edwards,‡ who gives a short detailed description, the letters C.M., placed at the end, signifying that the specimen was in the Paris Museum, but without data. I have failed to find any later mention of this highly interesting form, and it seems probable that the type until recently was the only specimen known.

In September, 1892, the writer and several other members of the Museum Staff, paid a visit to Port Stephens, and during the trip two examples were obtained from Mr. Jackson, one of the local residents. In October of the same year, the late Mr. T. Mulhall presented a specimen which was stated to have been obtained in Port Jackson. Since that date two other examples have been received, one from Port Jackson, and the other collected at Newcastle by Mr. J. Mitchell, who kindly presented it to the Trustees.

The first specimens were readily identified by the figure, with which they agree in most of the important characters. There are only two points of difference. The teeth forming the crenulate margin of the fifth joint of the second antenna are smaller and more numerous than represented in the figure; and the alternating bands of darker and lighter colour, indicated on the legs, are much less distinct, or wanting, in our examples. Notwithstanding these differences, I think that our specimens are identical with the individual figured, which was probably collected at or near Port Jackson by some of the early voyagers.

---

* Latreille—Encyclopédie Methodique (atlas), pt. 24, 1818, pl. cccxx., fig. 2.
† Latreille—Encyclopédie Methodique (text), x., 1825, p. 416.
‡ Milne Edwards—Hist. Nat. Crust., ii., 1837, p. 283.

Cancellus typus, M. Edw., affords another instance of a similar kind. This species was described* without a definite habitat, but a few years ago it was found in South Australia, and shortly afterwards also in Port Jackson.†

The habits and exact zone inhabited by this *Scyllarus* are unknown. It is, however, probable that it lives close in shore, at a considerable depth below the tide line, a region that has been little explored, having a distinct fauna which has yielded many new or rare forms, a few of which may be here noted :—

Tropidostethus rhothophilus, Ogilby, a small fish of the family Atherinidæ, is common all the year round, and lives in the surf, being rarely seen except in the white foaming water. Notwithstanding its abundance, it remained unnoticed until found by the writer in 1893.

Hemiscyllium modestum, Günther, commonly called the Blind Shark, inhabits the rocky recesses immediately below the tide line, and can only be obtained by fishing with a hook and line.

A rare zoophyte, belonging to the genus *Myriothela*, and differing little from *M. phrygia*, Fabr., lives on seaweeds in the same zone, and until recently it had not been observed in Australian waters.

The most remarkable example, however, is a small tubicolous Amphipod, which I refer without hesitation to the genus *Siphonœcetes*. This genus, according to G. O. Sars, contains but three species—one arctic and two occurring on the coast of Norway. Early in the present year I found a number of examples of what I believe to be a fourth species, at Maroubra Bay, which had been washed up from a considerable depth during a heavy gale. The spot where the specimens were gathered has been my favourite collecting ground for many years, although I never met with the species before, and have since carefully searched the locality without avail. I am of opinion that this species lives in that intermediate zone which is out of reach from the shore at the lowest tides, and too rocky and inaccessible to the trawl or dredge. Occasionally at certain seasons, denizens of this region wander about the spring tide line, and may at such times be captured. As an instance, mention may be made of the rare and curious Isopod, *Amphoroidea australiensis*, Dana, a species of an olive-green colour like the plants upon which it feeds, and to which it clings so tenaciously that it can only be removed with difficulty. A single example was found on a loose piece of seaweed at Maroubra Bay in May, 1896. It was afterwards searched for in vain during my weekly visits, and was not met with again until December, 1898, when about fifty examples were obtained, and in January of the present year

---

* Milne-Edwards—Hist. Nat. Crust., ii., 1837, p. 243.
† See List of Invertebrate Fauna—Proc. Roy. Soc. N. S. W., xxiii., 1889, p. 232.

many more were secured by wading and pulling up the seaweeds from deep holes. The plants were carried up above high water-mark, and left for an hour or so to dry. The drying has the effect of loosening the hold of the Isopods, and they may be readily detached by shaking the plants.

The following is a complete description of *S. sculptus* :—

SCYLLARUS SCULPTUS, *Latreille.*

Scyllarus sculptus, Latr., Encyclopédie Methodique, pt. 24, 1818, pl. cccxx., fig. 2, and x., 1825, p. 416 ; Milne Edwards, Hist. Nat. Crust., ii., 1837, p. 283.

Adult male.—Carapace 90 mm. in length, the width of the front between the orbits 49 mm., that of the entire frontal margin (spines included) 81 mm., and that of the hinder margin 77 mm. ; the greatest width is in a transverse line with the posterior cardiac region.

The rostriform process is slightly depressed, somewhat emarginate anteriorly, with obtuse lateral angles ; its length is about 5 mm., its breadth anteriorly 6·5 mm., and its narrowest part is just above the base and measures 4 mm. It is bounded externally on each side by a transversely elongate C-shaped depression, into which the somital lobes are dove-tailed and appear to be fused with the carapace. Each lobe has about eight or nine rounded tubercles on the posterior border, a pair on the anterior externally, and a prominent wide based denticle about the middle ; its sides slope away to the rostrum and to the external pair of tubercles. The length of the lobe is 5 mm., and its width 10 mm.

Lateral frontal margins deflexed, slightly curved, the inner extremity ending in a low tooth, the base of which is situated under the external portion of the somital lobe ; the outer portion of the margin curves upwards and terminates at the rather pro-minent interno-orbital spine. The orbits are well defined ; the cavity beneath the cornea is margined with long hairs. The inner portion of the superior border is elevated and bears three prominent spines and as many low tubercles ; centrally there is a pair of tubercles bounded on each side by a sinus, and the outer portion has a series of four tubercles and a spine at the angle. Anterior to the latter is a wide sinus, and a large spine-like process which arises from a point near the insertion of the second antenna and forms the outer boundary of a large anterior orbital fissure ; the inner side is limited by descending process of the front. The base of the fissure is occupied by a lobe a little higher than broad at the base, and uni- or bituberculate at the summit.

The upper surface of the carapace bears a few more or less acute spines, and is closely covered with flattened scale-like tubercles ; each tubercle is fringed anteriorly with short stiff setæ, posteriorly the setæ are few or wanting. The spines are disposed as follows: three equally spaced in a median line, the first situated at the base of the rostrum, the second and third on the gastric region. There is a spine immediately posterior to each orbit, and another on each side, situated at a short distance inwards behind the latter and in a line with it and the antero-lateral angle of the carapace. There are five or six submedian pairs ; the first are seated on the gastric region and are rather widely separated, the remaining pairs are arranged in the form of a narrow V on the cardiac area, with a single median spine at their base ; a few occur on the branchial region in a line with the inner orbital prominence, and about ten form a transverse series at a short distance from the hinder margin of the carapace.

Lateral margins of the carapace armed with fourteen or fifteen more or less compressed spines ; of these, are in advance of the cervical incision, the anterior one is large and prominent, the succeeding four gradually diminish in size as the well-marked cervical groove is approached, the latter is bounded posteriorly by a spine equal to or larger than that at the antero-lateral angle, the eight following are subequal in size and in distance apart, their outer borders are beset with tubercles. An irregular sub-marginal series of spiniform tubercles is situated on the posterior half of the carapace immediately above the lateral margin ; these form one or two rows and are most pronounced posteriorly.

A sharply defined, deep, smoothish, transverse groove extends from side to side across the hinder part of the carapace, at a distance of about 7 mm. from the ciliated posterior margin ; laterally the groove is nearly twice as wide as at the centre.

Pleon strongly sculptured, clothed with setiferous tubercles. The first segment is evenly convex above and below, and exhibits superiorly a well-marked transverse groove, situated much nearer the posterior than the concealed anterior margin, and it descends on each side to the coxal plate; the latter is small and bilobed. The anterior lobe is broad, rounded, and its margin is continued inwards to the end of the transverse groove ; the posterior lobe is triangulate and acute, its point is directed outwards and down-wards ; the posterior margin is slightly crenate and oblique.

The second to sixth segments bear raised arborescent markings; all, except the last named, have a median carina. Each segment is transversely grooved and somewhat depressed anteriorly ; the groove is bounded by a raised crenated line, which is just visible on the first three, but cannot be seen on those succeeding without bending the pleon downwards. The inner process of each coxal plate is limited by a well defined smooth groove, which extends from the anterior depression to the articular condyle.

The inferior surface of the second segment is much less curved than the first ; the convexity diminishes to the fifth, which is almost straight; all except the sixth and seventh have a prominent transverse denticulated ridge, which is directed backwards ; it is centrally situated in the first and second. In the succeeding three the posterior margin to which the connecting membrane is attached is very short, and is overlapped by the denticulate ridge. In the female this ridge is present, but it is smooth and much less distinct in all except the first segment.

The inferior surface of the sixth segment is flat, punctate, smooth, with a slight longitudinal median groove, it is imperfectly calcified, and scarcely higher than the bounding membrane. In length it exceeds the two preceding, and is equal to the third ; the latter is shorter than the second or first, which are twice as long as the fourth.

Superiorly the segments, except the first and seventh, are sub-equal in length. Their width gradually diminishes; the first measures 54 mm. across between the joints, and the fifth 35 mm.

The surface ornamentation consists of groups of arborescent patterns, and flat subimbricated scales ; the latter occur chiefly on the upper and on the lower posterior regions of the coxal plates. The former are present on the second to the fifth segments and disposed as follows : a narrow median group which forms the longitudinal keel ; on the second the keel commences about the middle and is continued to near the posterior margin, where it becomes transverse and extends nearly the whole width of the segment. There are two prominent groups, one on each side of the keel, but separated from it by a longitudinal row of two or more rounded tubercles.

On the remaining somites the dendritic sculpture becomes a little less distinct, but the keel is longer, and on the fifth it equals the segment.

The coxal plate of the second segment is very large and bilobed. It bears a Y-shaped impression on its surface, and from twelve to fourteen spines or tubercles along its margin. The front portion of the margin is transverse, the outer is obliquely directed backwards and downwards, and terminates in a large triangular tooth. The margin of the posterior lobe has five rather large denticles, and is slightly curved forwards towards its extremity.

The plates on the next four segments agree in having the anterior margin elongate, straight, almost smooth, and slightly crenate, and the posterior more or less curved and dentate like the second. The outer border, however, differs in each. In the third it is obliquely truncated and obscurely bi- or tridentate, the posterior denticle being rather large. The fourth is truncate and may be regarded as one broad lobe or tooth. The fifth and sixth are triangulate, the latter somewhat elongate.

The sixth segment has superiorly a faint longitudinal mesial groove, and two pairs of submedian tubercles ; the anterior pair are separated by the groove, the posterior pair are much wider apart and nearer the hinder margin ; the latter bear about eight tubercles, four on each side of a short central ridge.

The uropods and telson have the basal third calcified. The superior distal margin of the outer ramus has eight small denticles, and that of the inner ramus two or three situated on the outer fourth.

The telson is marked by four smooth pit-like depressions, the anterior pair are wide apart and seated on the base, the posterior pair are large, submedian, and occupy the distal half of the calcified part. The membraneous laminæ of the uropods and telson have numerous radiating ribs, which exhibit repeated dichotomous branching ; the ribs on both surfaces of the telson and on the outer halves of the rami are closely granulose.

The pleopoda are biramous ; the first pair are wanting in both male and female. The second pair in the former are foliaceous and have the rami equal in length ; the outer one is broad at the base, and the inner one rather narrow. In the succeeding pairs the outer ramus is similar in shape to that of the second, but the rami decrease in size on each somite ; the inner ramus is represented by a short obtuse conical stump. It is interesting to note that the inner ramus of the fourth segment on the left side—in the male example under notice—is considerably developed, it is equal to the outer in length but is very narrow and styliform ; a small stylamblys occurs at its base.

The second pair of pleopoda in the female are very large, and consist of foliaceus plates with strongly ciliated margins. The inner ramus is rather broader than the outer, and carries a single stylamblys tipped with long setæ, to which the ova are attached. In the following pairs the lamina of the outer rami is much smaller than those of the second ; the inner ramus is narrow, elongate, and three-jointed ; a foliate plate arises from the outer base of the second joint. The margins of the joints and lamina are more or less fringed with long plumose hairs to which the bundles of ova are cemented.

Basal joint of first antenna tuberculate, twice as long as broad, the inner distal angle terminating in a prominent denticle ; second joint stout, smooth, compressed, nearly twice as long as the first, and equal to the third ; fourth shorter than the latter, about twice as long as the outer flagellum.

Outer antennæ a little shorter than the inner. The basal joint is very large ; the inferior distal margin is denticulate, it carries about eighteen denticles, including the larger ones situated at the inner and outer angles. Second joint twice as broad as long ; the superior distal margin bears five spines ; of these, two are situated on the outer fourth, one about the middle and two on a produced lobe, which projects in front of the basal joint of the first antennæ and partly conceals the lower third of the second joint.

Third joint greatly expanded, somewhat cordiform in outline ; the outer margin is neatly curved, and bears six teeth, each of which is minutely denticulated. The distal margin is oblique, and has four or five large teeth ; the inner border is tridentate, the denticules occur on a projecting lobe similar to, but larger, than that on the inner angle of the preceding joint.

Fourth joint as broad as long, the inner border with three spines and the distal margin with two or more ; the outer distal angle is produced on the lower surface, but not on the upper.

Fifth joint nearly one-fourth broader than long ; its outline is semi-elliptical ; the margin is shortly ciliate, the inner third is acutely dentate, the rest crenate, the surfaces are minutely hairy punctate ; the upper surfaces of all the preceding joints are more or less scaly or tuberculate.

First pair of legs stout ; the inferior surfaces of the first three joints are beset with elevated tubercles, with dark brown tips; the merus is somewhat compressed, the inner surface is smooth and adapted to the carapace, the outer is slightly depressed longitudin- ally; near the upper border the latter is subcarinate, smooth, and shortly setose ; the inner lower border is similar to the upper, but is less elevated and only half its length ; the outer aspect of the lower border is marked by a series of ten subspiniform scales ; the distal margin has four or five obscure spines. Carpus half as long as the merus, with a strongly marked groove on the outer surface near the rounded upper border.

Propodus tapering, its greatest depth equal to half the length, its diameter equalling the depth at the distal extremity, the upper and lower borders are rounded. Tarsus curved, as long as its supporting joint, upper surface somewhat flattened, internal aspect grooved, the lower with two tufts of setæ near the base ; the distal half is horny, the corneus and calcareous portions overlap at their junction, the latter at the sides and the former above and below.

Second pair of legs the longest, their length without the tarsus equal to the first and also to the third ; fourth and fifth shorter, the latter nearly equal to the first without the tarsus.

The merus joint of the second pair is one-fourth longer than that of the first and of the fourth, about one-eighth longer than the third, and nearly twice the length of that of the fifth ; it exhibits—like the first and those succeeding—a shallow longitudinal depression near the upper border. The carpal joints are subequal, and each has a well-marked longitudinal groove on the external surface. The propodal joints also exhibit a shallow groove externally, and a line of short setæ indicating another less distinct groove on the inner surface. The length of the joint varies slightly; the second measures 37 mm., the third and fourth 28 mm., and the fifth 32 mm.

The tarsal joints of the last three pairs are shorter and more curved than the first or second.

The basos of the fifth legs have the superior distal angles produced, the anterior bears a triangular tooth, and the posterior a crest-like lobe directed outwards and tipped with from three to five denticles; the lobe measures 6·5 mm. in length, and 5 mm. in height.

The ground colour is greyish-yellow, in parts almost obliterated by crimson tints; central area of fifth joints of the outer antennæ and the margins of the third joints wholly of this tint; the lateral borders and the posterior margins of the carapace, and pleon also, red. Meral joints of legs with a central transverse crimson band, extremities of legs purple or blue. Membrane of telson and uropods yellow, mottled with purple. Inferior surface of antennæ yellow dotted with red. The legs red, dotted with yellow.

Total length of body, 300 mm.

REGALECUS GLESNE, ASCANIUS.—AN ADDITION TO THE FAUNA OF NEW SOUTH WALES.

By EDGAR R. WAITE, F.L.S., Zoologist.

IN his article—"On a species of *Regalecus* or Great Oar-fish, caught in Okain's Bay,"[*] H. O. Forbes mentions that the example there recorded is the tenth known from New Zealand waters, the records being as follows:—(1) Nelson, Oct. 1860; (2) Jackson's Bay, 1874; (3) New Brighton, May 1876; (4) Little Waimangaroa Beach, 1877; (5) Cape Farewell, 1877; (6) Moeraki, 1881; (7) Moeraki, June 1883; (8) Otago, June 1887; (9) Nelson Harbour, September 1890; (10) Okain's Bay, May 1891.

Two examples have been recorded from Australian seas. McCoy[†] figures and describes under the name *Regalecus banksi*, Cuvier, a specimen taken in May 1878, between Victoria and Tasmania. The second occurrence was near the Tweed River, in Queensland waters, and C. W. De Vis publishes a description of the fish under the new name *Regalecus mastersii*.[‡] In his Catalogue[§] Macleay includes *Regalecus gladius*, Cuv. & Val., with the remark, "said to have been seen on the Tasmanian Coast. Johnston, possibly on this authority, places it in his list of Tasmanian species.[||] The example described by Ramsay as *Regalecus jacksoniensis*[¶] is really a *Trachypterus*.

I have now to record a third Australian example. On August 12th last, a very fine specimen was discovered by a fisherman cast up on Shark Beach, within Port Jackson. It was obtained by Mr. Fitzwilliam Wentworth, of Greycliff, Vaucluse, who presented it to the Trustees.

Examples of the genus have never been obtained perfect, due to the extreme fragility of their members. The portions missing in our specimen were the lower parts of the head, including the branchiostegals, the cephalic and nuchal crests, one of the ventral filaments, and the end of the caudal. The body was cut almost in halves, by, it is presumed, a boat: otherwise it was in fair

---

* Forbes—Trans. New Zeal. Inst., xxiv., 1891, p. 192.
† McCoy—Prod. Zool. Vict., Dec. xv., 1887, pl. 145.
‡ De Vis—Proc. Roy. Soc. Qd., viii., 1892, p. 109.
§ Macleay—Proc. Linn. Soc. N.S.W., vi., 1882, p. 55.
|| Johnston—Proc. Roy. Soc. Tasmania, 1890, p. 34.
¶ Ramsay—Proc. Linn. Soc. N.S.W., v., 1881, p. 631, pl. xx.

condition and the appended notes were made while the taxidermists were engaged upon their work, no time having to be lost in its preparation. It proved to be a female, as with every other specimen of *Regalecus* sexually examined.

REGALECUS GLESNE, *Ascanius.*

D. 340 + ?   V. 1.   P. 12.

Height of body 15·46, length of head 20·1 in the total length· Eye 5·0 and maxilla 28·5 in the length of the head. Teeth absent· The head and body together are much shorter than the tail, being contained therein 3·5 times. The gill rakers of the first arch are very long, the longest measuring 28 mm.; they are slender and tapering and are furnished with short stiff hairs on their inner margins; there are five rakers on the hinder and thirty-four on the lower limb—there is also a very small raker between each of the larger ones, it may easily be overlooked. A second set occurs on the inner side of the arch, these rakers are comparatively small and are arranged in three series, the largest not more than 6 mm. in length. The first series consists of a rather broad triangular plate, surmounted by a thickened head which bears a number of hairs. The second series is alternate with the first one, and the rakers are similar, but only half their size. The third series is formed of rakers less than half the size of those of the second series, they occur between each of the others and are therefore twice as numerous, these are comparable with the small alternate rakers described on the outer set of the first arch. The rakers on the other arches (there are four and a half in all) are very small and are similar to the inner set of the first arch. The tail is possibly incomplete, but I do not think very much has been lost, the membrane extends to its tip. Of the cephalic rays the first is the only one now perfect, it is very stout basally but tapers to a thread, it measures two and a quarter times the length of the head; the four following are slender, close together but are broken off at about one-third the length of the first one, the others are broken short, all were connected by membrane. The dorsal fin is tolerably perfect but broken away at intervals, its height is one-half more than the diameter of the eye. The only pectoral fin remaining is broken, but it has not the vertical aspect ascribed to some forms, and exhibits the bases of twelve rays. One of the ventral filaments is complete and extends nearly to the vent, it terminates in a fleshy tag. The lateral line arises above the operculum, passes obliquely downwards and runs along the lower half of the body at about one-fourth its height from the ventral surface.

The skin is very thin and is marked by five longitudinal rows of tubercles, where these tubercles exist the skin is attached to the flesh beneath, but the spaces between are quite free and a pencil or penholder may be passed under the skin along the entire length of the body.

Colour—The general colour is silvery-grey, the upper surface is quite silvery and the five tubercular rows are equally so. The body is marked with irregularly oval black spots, those in front being as large as half a crown, while those towards the end of the tail are much smaller—about the size of a shilling piece. On the lower surface the spots are more widely spaced, and narrow black vertical bars are placed at intervals and directed obliquely from before backwards. The fins including the ventral filaments, are flaming red ; the dorsal is variegated with a series of opal coloured spots, one to each ray, placed nearer the body than the edge of the fin; mouth within black. The vertebrae are 136 + ? in number.

Dimensions.

Total length...	4020	mm.
Length of head	200	,,
Height do.	190	,,
Diameter of eye	40	,,
Length of maxilla ...	70	,,
Height of body	260	,,
Head to vent	1150	,,
Length, cephalic filament ...	450	,,
Ventral filament, including tag	280	,,
Ventral terminal tag	60	,,
Height of dorsal fin	65	,,

The general inclination is to now regard all the described examples as referable to one species, *R. glesne*, and although this name is adopted for our example, it is to be borne in mind that specimens exhibit very great diversity in proportionate height and length. It has, on the other hand, been pointed out that the proportions of head to body and the number of fin rays differ greatly at different ages. Parker* has drawn up a key to the species of *Regalecus*, from this our specimen appears to be referable either to *R. banksii* or *R. grillii*, having more than two hundred and fifty dorsal rays, no teeth, and no caudal fin. In the former species the height is one-thirteenth of the length and the dorsal rays two hundred and seventy-six to three hundred and five. In the latter the height is one-eighteenth of the length and the dorsal rays four hundred and six. In McCoy's example, named *R. banksii*, the height is one-twenty-third of the length and the dorsal rays four hundred and six. De Vis' species *(R. mastersii)* is the least elongate known, its height being but one-ninth of its length ; the number of dorsal rays is not given.

---

* T. J. Parker—Trans. N. Zeal. Inst., xvi., 1883, p. 296.

OCCASIONAL NOTES.

III.—*PALÆOPEDE WHITELEGGEI*, ETH. FIL.

In the last number of this publication* I described a supposed Endophytic Alga, under the above name, occurring in a *Favosites* from the Middle Devonian Limestone of Moore Creek, near Tamworth, in the north-eastern part of this Colony. Since then, on examining a composite coral (not yet described), from a limestone, possibly of much the same age, at Goodravale, on the Goodradigbee River, not far from the junction of the latter with the Murrumbidgee River, I again met with a similar organism. As in the former instance the coral tissues are traversed by the alga in various directions, and the characters of the intruder are practically the same as those of *P. whiteleggei*. In the present instance the heterocysts (?) vary in diameter from ·01 to ·015 mm.; the diameter of the filaments of cells from ·006 to 008 mm.; the distance between the heterocysts (?) ·07 mm.; and the total length of a filament generally about ·2 mm.

It is very interesting to meet with this Endophytic Alga and obtain confirmatory evidence of its existence in a perfectly distinct coral and at two places so widely separated, although near about the same geological horizon.

R. ETHERIDGE, JUNR.

IV.—*LAMPRIS LUNA*, GMELIN.

ITS RECURRENCE IN NEW ZEALAND WATERS.

On the 25th instant (October), we received notice from the Fresh Food and Ice Co., of this city, that a rare fish would shortly arrive from New Zealand. With the Curator (Mr. R. Etheridge, Junr.) I visited the depôt of the Company, and there saw a very fine example of the Opah, *Lampris luna*. We learned that it had been so identified in New Zealand, and the daily papers of 26th October, contained the announcement that the fish exhibited in the window of the Company " was caught by some men employed on the Timaru Dredge, New Zealand, and after being exhibited in Timaru was forwarded to Dunedin. Professor Benham, of the Otago

---

* Etheridge—Rec. Austr. Mus., iii., 5, 1899, p. 127.

University, on seeing the fish, pronounced it to be *Lampris luna*, of which only one previous specimen is recorded as having been caught in New Zealand, namely on the coast of Wellington in 1883."

A short notice of this former occurrence was published by Hector,[*] who states that the superb fish was cast on the beach near the Manawatu River, in December 1882, and presented to the Wellington Museum.

In 1895 a specimen was obtained in Tasmania This is recorded by Morton,[†] who writes :—" This very interesting specimen, the first on record of having been captured in Australia, was forwarded to the Museum by Mr. Alfred Lovett, on December 18th, 1895. Mr. Lovett states that it was found washed up on the beach near Port Arthur, and weighed 130 lbs. Unfortunately the crows had picked and destroyed one side. . . . the branchiostegals were six in number, there was no sign of a seventh."

The specimen now in Sydney, has been exhibited for the past three weeks in the window of the Fresh Food and Ice Co., daily, and placed upon ice for preservation at night. Owing to the price at which the fish was valued by the Company, the Trustees of the Museum have not been able to secure it for the National Collection. The following dimensions have been kindly supplied by the Manager of the Company :—Length 3ft. 6in., height 2ft. 4in., thickness 9in.

EDGAR R. WAITE.

* Hector—Trans. New Zeal. Inst., xvi., 1883, p. 322.
† Morton—Proc. Roy. Soc. Tasm., 1896 (1897) p. 99.

TWO NEW THOMISIDS.

By W. J. RAINBOW, F.L.S., Entomologist.

(Plate xxx.)

THE present paper contains descriptions of two new species of Thomisidæ, exceedingly interesting forms. The first, for which I propose the name *Misumena tristania*, was collected by Mrs. C. T. Starkey, of Neutral Bay, Sydney, at Cobbity, and presented by her to the Trustees. The spider was found upon some flowers of *Tristania laurina*, R. Br. (N.O. Myrtaceæ) a tall shrub or small tree, which (according to Mr. J. H. Maiden, Director of the Botanical Gardens, Sydney, who kindly identified the flowers for me), is found near watercourses, and is not uncommon in mountain gullies. The flowers of this shrub are bright yellow, and the colour is closely simulated by the spider, so that, unless it happens to move, it is difficult to detect. And, to colour must be added the form of the animal itself, when studying it in connection with the question of mimicry, because, when at rest, the body is posed in such a manner as to add greatly to the effect. The mimetic resemblance, therefore, must be regarded as *protective* and *attractive*: the former, because it would assist the Arachnid in evading insectivorous foes, and the latter, because it would be the means of attracting insects upon which it feeds.

The second species, *Saccodomus formivorus*, is remarkable not alone on account of its form, but also for the reason that, contrary to all previously recorded facts based upon accurate observations of the habits of the Thomisidæ, it constructs a baglike nest. It is in consideration of this, that in founding the genus diagnosed below, I propose the generic term *Saccodomus*.

It has long been regarded as an axiom that all those individuals and species embraced within the vast family Thomisidæ, popularly known as "Crab" or "Flower" Spiders, are essentially wanderers, constructing neither webs nor nests for the capture of prey, but relying solely upon their activity, cunning, or mimetic resemblance to surrounding objects.

Another interesting feature in connection with this species is its food. Both Mr. W. W. Froggatt and Mr. George Masters, in speaking to me upon the subject, said that so far as their

A

observations went, these spiders lived entirely upon small ants—
Leptomyrmex erythrocephalus, Fab.   These insects were, Mr.
Froggatt says, drawn into the nests by the spiders, where they
sucked the vital juices from their bodies, and then threw them
out.   Mr. Masters, who collected some specimens and nests at
Glen Oak, Williams River, informed me that the nests were
covered with small ants, and that it was their great numbers on
the trees that attracted his attention.

Mr. D. A. Porter, of Tamworth, who has kindly donated both
spiders and nests to the Trustees, wrote me as follows :—" The
trap is a 'bag,' or *cul-de-sac,* with the opening upwards, and
generally fixed between the small branches [in a fork].   The ants
travel *over* and *on* it in going *upwards,* but often 'fall in,' probably when coming down."   Further, my correspondent says that
he has observed the ants inside, travelling round and endeavouring
to escape, but that he has never noticed one succeed, or try
to climb out ; he has also seen fragments of dead ants in the
traps.   Mr. Porter is, without doubt, an admirable observer,
and his remarks are therefore of value.   In his letter he says,
that although he had never noticed a spider in or upon one of
these traps, he had always found them not far away, and
as "they were always of the same kind, concluded that they
were the owners."   It is possible, however, that some of the
spiders were not at home, when Mr. Porter examined the nests,
or that they may have escaped his observation, because upon
closely examining the group of nests presented, I found a living
example hiding deep down in one of the bags.   The colour of the
animal's abdomen, and the web of which the nests are made, are
so much alike that detection is very difficult, and under some
conditions impossible.

Mr. Porter informs me that in the Tamworth District these
nests or traps, so far as he can say, are only found upon one
variety of bush, known to the aborigines as " Dthluby."   I am
again indebted to Mr. J. H. Maiden for his kindness in identifying
this species.   It is the common " Whitethorn," *Bursaria spinosa,*
Cav. (N.O. Pittosporæ).   From specimens I have handled, however, it is evident that this spider constructs its nest in various
trees or shrubs.   The one obtained by Mr. Froggatt at Aclong
was taken from one of the tea-trees *(Leptospermum sp.),* and that
by Mr. Masters from a Eucalypt.   Mr. Porter says in concluding
his note that the traps are generally situated at two to four feet
from the ground.*

---

* For description and figures of nests the reader is referred to a previous paper by me in Proc. Linn. Soc., N.S.W., xxii., 3, 1897, p. 549, pl.
xviii., figs. 6 and 6*a*.

Family THOMISIDÆ.

Sub-family MISUMENINÆ.

Genus Misumena, *Latr*

MISUMENA TRISTANIA, *sp. nov.*

(Pl. xxx., Figs. 1, 1*a*.)

♀. Cephalothorax 2·8 mm. long, 3 mm. broad ; abdomen 5·5 mm. long, 5·8 mm. broad.

Cephalothorax obovate, arched, broader than long, orange-brown relieved by chrome yellow and pitchy-black markings.   *Pars cephalica* arched, truncated in front, orange-brown, ornamented by a median line of chrome yellow and a few small concolorous spots; in addition to these there are also upon the upper surface two fine lateral chrome yellow lines : these commence well forward, curve gently first in an outward direction, and then more sharply inwards, ultimately meeting at the base ; on the outer side of each of these lines, and commencing at a point immediately below, but in a line with the posterior lateral eyes, there is a broad, wavy, longitudinal pitchy-black line, which does not terminate until near the centre of the cephalic segment ; these lines are narrowest in front, become gradually wider, and terminate in an obtuse point ; immediately below the anterior row of eyes (the *clypeus*) there is a broad, strongly recurved bar of chrome yellow ; from below each lateral eye of the anterior row, there is directed backwards and outwards, a narrow, wavy, concolorous line, and immediately below the posterior median eyes, there is a rather sharp depression or pit which is broadest in front, and has its margins chrome yellow.   *Pars thoracica* arched, broad, orange-brown with a large triangular patch of chrome yellow at junction of cephalic and thoracic segments, and enclosed between the pitchy-black lines referred to above.   *Marginal band* chrome yellow, broad.

Eyes small, black, normal.

Legs orange-brown, extremities of tibiæ of first and second pairs nearly encircled with a deep, black band ; extremities of each haunch, trochanter, femur, patella, and tibia encircled with a band of chrome yellow ; femurs, tibiæ, and metatarsi clothed with fine adpressed hairs upon their upper surface, and armed with small lateral spines ; tarsi hairy ; tarsal claws black.   Relative lengths 1, 2, 4, 3.

Palpi short, strong, orange-brown, clothed with fine black hairs, and terminating with a small black spine.

Falces strong, convex, orange-brown, sparingly hairy.

Maxillæ yellow, moderately long, convex, slightly constricted at their centre, sparingly pubescent, apices inclining inwards.

Labium concolorous, rather longer than broad, truncated at tip.

Sternum yellow, glossy, convex, shield-shaped, sparingly clothed with rather long, strong hairs.

Abdomen broadly obovate, strongly arched, slightly projecting over base of cephalothorax, very finely pubescent, chrome yellow, relieved towards anterior extremity and sides with a series of black and dark brown markings, and irregularly shaped large and small concolorous spots; in addition to these there is also present, and running down the centre, a delicate net-work of tracery, barely visible to the naked eye; the inferior surface is also finely pubescent, chrome yellow and ornamented down the centre with a series of twelve dark brown spots, arranged in pairs, rather widely apart, the posterior pair especially so.

Epigyne as in figure.

Hab. Cobbitty, New South Wales.

Genus Saccodomus,* gen. nov.

The species for which a new genus is now proposed was first brought under my notice some years ago by Mr. W. W. Froggatt, who gave me both the specimen and its nest. These have since been added to the collection of the Australian Museum. When examining it at the time it appeared to me there could be little doubt as to the position it should occupy, namely, in the sub-family Misumeninæ. But, whilst many features appeared to point to this sub-family as its correct place, there were nevertheless, some that were decidedly conflicting, and these suggested certain important analogies with the sub family Stephanopsinæ. Indeed, the form was so novel, that I hesitated to describe it until I could obtain more specimens, and devote more time and labour to the elucidation of the problems presented.

In my studies I was, happily, assisted by my esteemed friend, Mr. H. R. Hogg, M.A., of Melbourne, who, writing to me upon the subject said :—"In this group [*i e.*, Misumeneæ] it has the nearest affinity with the sub-group Diæcæ, and I place it as a new genus between *Heriæus*, E. Simon, and *Diæa*, Thorell, the chief points keeping it out of either being : forehead sloping instead of vertical ; legs smooth instead of having certain bespinements ; rear row of eyes less recurved than front row."

During the interval that has elapsed since the above was written the subject has been further studied, with the result that both Mr. Hogg and myself still incline to the opinion that this species should form a new genus to be placed—provisionally—between *Heriæus* and *Diæa*.

Nevertheless there are yet, to my mind, some very debateable points to be removed before the subject can be satisfactorily settled, and it may even be necessary hereafter, either to amend one of the existing sub-families, or to found a new one for its

* Derivation : Σάκκος, a bag or purse; δόμος a dwelling place.

reception, and by way of illustration the following tables are submitted :—

Points showing. wherein the genus *Saccodomus* differs from genera included in the sub-families Stephanopsinæ and Misumeninæ :

Sub-family Stephanopsinæ.	Genus *Saccodomus*.
First pair of legs longer than second;	First and second pairs equal;
Maxillæ parallel;	Maxillæ slightly inclined inwards;
Forehead vertical;	Forehead sloping;
Both rows of eyes equally recurved;	Posterior row of eyes less recurved than anterior row;
Teeth on lower margin of falx;	Lower margin of falx smooth;
No row of hairs upon exterior margin of falx.	Hairs present upon exterior margin of falx.

Sub-family Misumeninæ.	Genus *Saccodomus*.
Second pair of legs longest;	First and second pairs equal;
Forehead square, vertical;	Forehead square, sloping;
Both rows of eyes equally recurved.	Posterior row of eyes less recurved than anterior row.

Points in which the genus *Saccodomus* agree with—

Sub-family Stephanopsinæ.	Sub-family Misumeninæ.
Labium (with some genera);	Maxillæ inclining inwards;
Forehead not attenuated;	Shape of labium;
Front femur not bespined;	Lower edge of falx smooth;
Anterior row of eyes not near together.	Tarsi without claw tufts;
	Front row of eyes;
	Hairs upon clypeus;
	Skin bristly;
	Forehead square;
	No spines on tarsi or metatarsi.

It will be seen, from the above comparative tables, that the majority of points rests in favour of the sub-family Misumeninæ, in which for the present it is placed; and again, if the reader will refer to pl. xxx., figs. 2 and 2*b*, he will note, so far as the abdomen is concerned, a striking resemblance in contour to those species forming the sub-family Stophanopsinæ.

Until quite recently our collection contained only one specimen of this remarkable spider, and this fact made me dubious about describing it. Fortunately, however, an esteemed correspondent, Mr. D. A. Porter, of Tamworth, to whom reference has already been made, forwarded additional specimens to the Trustees, by the aid of which supplementary material, I feel I am now justified not only in describing the species, but also in founding a new genus for its reception.

CHARACTERS OF GENUS.

Cephalothorax longer than wide, high ; sides and posterior angle sharply declivous. *Pars cephalica* sloping sharply forward, broad, square in front, and truncated. *Clypeus* hairy. *Pars thoracica* high, sides declivous, and deeply indented laterally.

Eyes small ; posterior row less recurved than the anterior ; of the four comprising the anterior row, the median pair is smallest and much the closest together, whilst the series constituting the posterior row, are widely separated from each other, and equidistant.

Legs long, hairy.   Relative lengths $1 = 2, 4, 3.$

Palpi short, hairy.

Falces robust, hairy.

Maxillæ moderately long, outer angles constricted near their centre ; apices inclining inwards, obtuse, and not divergent.

Labium coniform.

Sternum oval, convex, truncated in front, obtuse behind.

Abdomen oval.

SACCODOMUS FORMIVORUS, *sp. nov.*
(Plate xxx., figs. 2, 2*a*, 2*b*, 2*c*, 2*d*, 2*e*.)

♀ . Cephalothorax 2·7 mm. long, 2·2 mm. wide ; abdomen, 4 mm. long, 3·7 mm. wide.

Cephalothorax longer than wide, high, sparingly clothed with fine, hoary pubescence, dark brown, encircled with exception of clypeus, by a deep cream-coloured band. *Pars cephalica* dark brown, sloping forward, truncated in front, sides declivous. *Clypeus* dark brown, hairy. *Pars thoracica* dark brown, high, sloping forward, sides and posterior angle sharply declivous; there is also on each side and near the junction of the cephalic and thoracic segments, a long, deep depression. *Marginal band* broad, cream-coloured.

Eyes small, black, arranged in two recurved rows of four each; second row less recurved than the anterior; of the four comprising the anterior series, the median pair is the smallest, much the closest together and separated from each other by a space equal to about four times their individual diameter, but each lateral eye is removed from its neighbour by many times its individual diameter; the posterior series are slightly larger than their anterior lateral neighbours, widely separated from each other, but equidistant or nearly so.

Legs robust, hairy; anterior pairs dark brown, joints annulated white, but, judging from the specimens before me, these limbs are subject to a slight variation in colour; posterior pairs pale yellowish, with dark brown markings. Relative lengths $1 = 2, 4, 3$.

Palpi short, dark brown, hairy.

Falces robust, concolorous, clothed with short stiff hairs.

Maxillæ dark brown above, outer angles yellowish, moderately long, arched, inclining inwards, apices obtuse, surface clothed with short, stiff hairs.

Labium dark brown, coniform, convex, clothed with coarse greyish hairs.

Sternum greyish, oval, moderately convex, truncated in front, obtuse behind, clothed with coarse greyish hairs.

Abdomen oval, moderately arched, truncated in front, projecting over base of cephalothorax; anterior, lateral, and posterior angles furrowed; superior surface and sides sparingly clothed with short black hairs; near the posterior angle, and in the median line, there is a rather deep puncture; below this, and at about one-third the length of the abdomen, there is a row of two punctures, equally as deep and large as the one referred to above, but widely separated from each other; below these, and just beyond the centre, there is another row of two, but these are much larger than the preceding, and rather wider apart; a little below the second pair there is a recurved row of punctures, the lateral individuals of which are smaller than those already described, whilst the intermediate series (six) is much smaller still; below this again there is another recurved row of six small punctures; each of the recurved rows here described are seated in two transverse furrows; lateral angles furrowed longitudinally and finely punctured; dorsal, ventral, and lateral surfaces dull yellowish.

Epigyne as in figure.

Obs.—In gravid specimens the abdomen assumes a somewhat spherical form, the furrows and small punctures are entirely absent, whilst the deep, dorsal punctures so prominent in normal examples, are only barely visible.

Hab. Aelong (W. W. Froggatt); Williams River (G. Masters); Tamworth (D. A. Porter).

SPEARS WITH INCISED ORNAMENT.

By R. ETHERIDGE, Junr., Curator.

In 1897, I fully described* an Australian Spear with incised ornament, extending nearly the whole length of the weapon. I further commented on the rarity of this form of sculpture amongst Aboriginal spears, and the general absence of illustrations in works of reference. In working through the store collection of Ethnology, I met with four additional examples, so far similar, that nearly the entire surfaces are covered with ornamental incisions, but all differing in the motive, and three of them greatly so, from the spear referred to.

The first is of the same length as the already described specimen, viz., eleven feet nine inches, but lacks the colour bands near the point or apex. The serpentine longitudinal grooves are five instead of six in number, and extend from within two feet three inches of the point, and nine inches of the butt. The grooves are toothed in a similar manner, but instead of the serrations looking backwards, i.e., towards the butt, they are presented forwards towards the point of the weapon ; furthermore, the interstices between the serpentine grooves are occupied by V-shaped and bird's feet (" broad arrow ") incisions, or simple oblique nicks, arranged with a certain degree of order, like with like. The ornament at the butt is finished off by four feather-like incised tags, consisting of a central groove margined by oblique nicks, reminding one of the feathered shaft of an arrow. This spear is said to have come from the Paroo River District, but from which side of the border I am unable to say.

The three remaining spears are much shorter weapons, being each a trifle over seven feet in length. On the first the sculpture is spiral, consisting of two bands, extending from the butt to within two feet of the point. The bands are each defined by two grooves, the interspaces being cross incised, thus giving them greater prominence and effect, but the apical two feet is variously occupied. First, proceeding upwards from the termination of the spiral bands, a rude representation of the human form is seen, with one arm only, and above this an oval body, both infilled with cross incisions. These are succeeded by sundry serpentine and meandering bands similar to the spiral already described. The interspaces are occupied by V-shaped incisions, oblique nicks, and what not.

---

* Rec. Aust. Mus., iii., 1, 1897, p. 6.

The second shorter spear is carved to a certain extent like the first. From the butt to within two feet seven inches of the point, two encircling or spiral bands traverse the surface exactly as in the first spear, but they are connected by two other bands running somewhat obliquely to the weapon's length, and at the same time are discontinuous, leaving free or unoccupied gaps. The result is that even these disconnected bands become in the long run elongately spiral in a contrary direction to the main bands, and, where present, divide the spear surface into long ovals. Some of the latter are occupied by zig-zag lines of nicks, either transverse to the length of the weapon, or oblique to it. The apical space of two feet seven inches contains an undoubted human figure, with both arms raised straight above the head, a boomerang-like object, and a third outline that may be intended to represent a fish ; the first and last figures are obliquely cross-incised. It is to be noted that the position of the arms in the human figure is one frequently seen in similar representations amongst the rock-carvings of the Sydney District, and on the dilly-baskets of North-Central Australia. Above the figures, and to the apex, the surface is occupied by a single broad spiral band cross-barred, the interstitial surfaces being ornamented in a similar manner to those of the lower portion of the weapon.

The third spear presents a complex style of incised sculpture, extending from within eight inches of the base to four inches of the point. It consists of short spiral bands terminating simply; others meander and return on themselves, either at one or both ends ; some again cross others forming oval loops by their intersection, ultimately becoming so complex that it is difficult to follow the pattern. Near the centre of the spear, on two of the interspaces are two objects that may be intended for shields, whilst on a third is another that has some general resemblance to a conventionalised bird. The human form is absent. The carving on this weapon is much rougher, and less well executed than on the others, In all, the section is circular, the ends pointed, acutely at the apex, obtusely at the butt.

I am unable to state, either the immediate locality of these spears, or the site of their manufacture. The first described by me was derived from Angledool, on the Narran River, close to the Queensland Border, in Central North New South Wales. The equally long weapon, now described, is believed to be from the Paroo River, rather more to the west, but from which side of the border is not known. It seems possible that whether manufactured or merely localised, this type of spear may be regarded as characteristic of the district in question. It is, however, very difficult and even hazardous, in the absence of definite information, added to the practice of barter, so common amongst the Australian Aborigines, to fix the locality of any weapon or implement.

LITTLE-KNOWN AND UNDESCRIBED PERMO-CAR-
BONIFEROUS PELECYPODA IN THE AUSTRALIAN
MUSEUM.

By R. ETHERIDGE, Junr., Curator.

(Plates xxxi. – xxxiii.)

Genus STUTCHBURIA,* *gen. nov.*

In our Permo-Carboniferous formation are two bivalves that have been variously referred to *Orthonota* by Morris, *Cardinia* by Dana, and one of them to *Pleurophorus* by DeKoninck, the determination of the last named author having been at various times accepted by myself and others; possibly also one or more of the shells from the same series of rocks, termed *Cypricardia* by Dana, may be congeneric. I have, however, for some time past, from the edentulous nature of the shells in question, doubted the propriety of these references.

. The species are *Orthonota?* *costata,* Morris (= *Pleurophorus morrisii,* DeKon.), and *O.?* *compressa,* Morris, which may, or may not be only the internal cast of *O. costata.* To these may perhaps be added *Pleurophorus biplex,* DeKon., and *P. randsi,* mihi. The internal structure of the two first, and particularly of *O.?* *costata* is known to some extent, but that of the third very little, and of the fourth not all. It is by no means certain that *P. biplex,* and *P. randsi* are congeneric with *O? costata* and *O? compressa,* and in consequence are left for the present in *Pleurophorus.* At the same time there is still an undescribed form in our Marine Series, that appears to be generically identical with *Pleurophorus:* this will be described later.

In form *O.?* *costata* and *O.?* *compressa* are narrow, transversely elongate, and more or less compressed Molluscs, inequilateral in the extreme, with simple pallial lines, strongly marked muscular scars, particularly the anterior, which are complex, and, so far as I can ascertain, edentulous, at any rate the examination of a very large number of internal casts has failed to reveal the presence of hinge teeth. In the place of the latter the cardinal margins were very much thickened, particularly at the extremities, and in all probability this was accompanied by an internal ligament. In the face of these combined characters the reference of the species

---

* Named in honour of Samuel Stutchbury, the first Government Geologist of New South Wales, as it then was.

in question to either *Cardinia*, *Orthonota*, or *Pleurophorus* appears to be impossible. It is a remarkable fact that the authors who have dealt with these shells invariably describe the cardinal margins as linear, narrow and concave. Indeed the remarks of both Dana and Morris indicate their mental uncertainty as to what genus they should be referred to.

I therefore propose, under the circumstances, the genus *Stutchburia* for the reception of *Orthonota*? *costata*, Morris, and if differing from it *O.*? *compressa* also, in honour of Samuel Stutchbury, the pioneer Naturalist, and one of the two pioneer Geologists of Australia.

The characters of the new genus will be as follows :—

Shell transversely elongate, equivalve, very inequilateral, the posterior end the longer, more or less compressed, closed, test thin; posterior slopes always rounded; a mesial sulcus sometimes present in each valve ; edentulous ; ligament supported on the thickened hinge plates ; dorsal or cardinal margins erect and sharp; umbones very anterior ; the anterior adductors large, with single smaller supplementary scars (?) between them and the umbones, in the cavity of which there are at times other scars ; posterior adductor scars large, but less defined ; pallial lines simple ; sculpture concentric and at times radiate.

The form, edentulous nature of the thin shell, internal ligament, and often radiate sculpture indicate the Solemyidæ as the family to which *Stutchburia* should be referred. The representatives of this family are *Solemya*, *Janeia*, and *Clinopistha*, to which Mr. W. H. Dall has suggested[*] the addition of *Orthodesma* and *Whitevesia*. Now, the proposed new genus, although resembling *Solemya* in its edentulous nature, and simple pallial line, differs entirely in having the ligament practically posterior, and no trace of the umbonal ligamental clefts. From *Janeia* it is easily distinguished by the equality of its valves, and from *Clinopistha* by outline, the presence of an internal ligament, and by the fact that the umbones are anterior and not posterior. The reference of *Orthodesma*, as described by Hall and Whitfield, to the Solemyidæ does not appear to be well established, but two of the species so described by Ulrich, from the Lower Silurian of Minnesota approach much nearer to *Stutchburia*, especially in their muscular scars.

With regard to *Whitevesia*, the edentulous nature of the hinge, simple pallial line, and internal ligament, indicate a departure towards our shell, but the grooved hinge plate, and both external and internal ligament if present, but of which there seems to be some doubt, as well as the very much feebler muscular scars,

_____

* Dall—Trans. Wagner Free Inst., iii., 1895, p. 515.

should be sufficient to separate them. I have seen no trace of an external ligament in actual specimens of *Stutchburia*, but in one of Morris' figures† of *S. costata*, as it must now be called, there is some shading above the hinge line that certainly does present the appearance of a cartilage, but I think it is misleading and not structural.

Before proceeding with the specific descriptions, a few generic points may be dwelt on more in detail. The dorsal or cardinal margins, or hinge lines of the valves, are erect and unquestionably closed, but on their inner and lower sides form thickened obtusely rounded edges. These continue past their conspicuous umbones, and in their substance immediately anterior to the latter, are excavated two depressions, one in each valve, which, in all probability gave attachment to the ligament at this end of the shell. If not of this nature, the only other solution is that these depressions are muscular. In the cast these thickened internal margins are represented by wide, shallow, longitudinal concavities, whilst the depressions are indicated by two sharp projections about midway between the umbones and the anterior adductor scars. The latter are large and deep, in the type species at any rate, and must have received strong and well developed muscles. On their posterior sides the interiors of the valves were much thickened, and in consequence deep depressions are left on the surface of casts, circumscribing the impressions of the muscles, which stand out boldly from the general surface, with an oblique inclination to the anterior.

The posterior adductor scars are situated high up on the flanks of the valves, immediately under the hinge lines, and although conspicuous, are less so than the anterior. They have an oblique inclination to the posterior, with the test correspondingly thickened on their anterior sides, but to a smaller extent than those of the other extremities of the shell. The simple pallial scars are well defined, continuous, and from their prominence in casts must have presented deep and sharp lines on the valve interiors.

The following are the species known to me :

STUTCHBURIA COSTATA, *Morris*, sp.

(Pl. xxxi., fig. 1.)

Orthonota ? *costata*, Morris, Strzelecki's Phys. Descrip. N.S. Wales, &c., 1845, p. 273, pl. 11, f. 1 (? excl. f. 2).

Cardinia ? *costata*, Dana, Wilkes' U. S. Explor. Exped., x., 1849, p. 692 (? pl. 4, f. 8, 8*a*, *b*, *c*.).

Pleurophorus Morrisii, DeKoninck, Pal. Foss. Nouv.-Galles du Sud, 3, 1877, p. 143, pl. 20, f. 5.

Pleurophorus Morrisii, Eth. fil., Cat. Austr. Foss., 1878, p. 77.

---

† Strzelecki.—Phys. Descrip. N. S. Wales, &c., 1845, pl. 11, f. 1.

(Compare *Cypricardia (Avicula?) veneris*, Dana, loc. cit., pl. 9, f. 3, 3*a* and *b*.)

Sp. Char.—Shell more or less compressed pod- or filbert-shaped, very inequilateral; dorsal and ventral margins almost parallel, the former straight, and the latter but little curved, with a slight inflection at about its anterior third; no diagonal ridges, but the valves uniformly and very slightly convex throughout, except for shallow cinctures running obliquely from the umbones posteriorly to the ventral margins; umbones inconspicuous, depressed. Anterior ends very small, slightly protruding, the margins rounded; posterior ends compressed, the margins obliquely rounded above and below. Ligamental fulcra large, leaving deep impressions in casts; ligamental (?) pits transversely elongated, inclined to triangular, represented in casts by sharp crests. Anterior adductor impressions very large, rather deltoid, concentrically grooved, the thickened posterior edges leaving wide groove-like depressions in casts, which extend to immediately in front of the umbones. Sixteen to twenty radiating costæ proceed from the umbones to the posterior margins commencing just behind the shallow oblique cinctures, with the whole surface crossed by close concentric fine lines, which imbricate the costæ.

Obs.—As this was the first species described by Morris, it must be regarded as the type. His fig. 1, of the reference quoted above, gives a faithful and accurate representation of the shell, and it will be observed that the radiating costæ are there visible on the cast, the test having broken away along the hinge line in both valves. Fig. 2 is an equally good illustration, but on this, although again a cast, the costæ are not visible at all. In DeKoninck's figure of this species the umbones are too acute and projecting, as they do not in reality overhang the anterior ends.

There appears to be every probability of Dana's *Cypricardia (Avicula?) veneris* being nothing more than a small individual of this species.

Loc. and Hor.—Jamberoo, Black Head, and Crooked River, near Gerringong, Illawarra District—Upper Marine Series.

STUTCHBURIA COMPRESSA, *Morris*, sp.

(Pl. xxxi., fig. 2, and xxxiii., fig. 1.)

Orthonota ? costata, Morris, loc. cit., pl. 11, fig. 2, (non f. 1)
Orthonota ? compressa, Morris, loc. cit., p. 274, pl. 13, f. 4.
Orthonota compressa, Eth. fil., loc. cit., p. 74.

Obs.—The specific value of this form mainly depends on the presumed absence of the posterior radiate sculpture, but as this is seen on one of Morris' types of *S. costata*, and not on the other, it may be taken as a specific character. I have therefore included the figure without the posterior radii as a synonym of *S. compressa*, and from the examination of several actual specimens, I believe the separation will hold good. It is also necessary to make the same remark on Dana's figure; he certainly describes *S. costata* plainly enough, but his illustration represents the form or condition known as *S. compressa* without a doubt.

The characters of *S. compressa* are practically those of *S. costata*, with the following exceptions :—The shell is rather more compressed, ligamentary pits of the hinge larger, anterior adductor scars subdivided by a groove, posterior adductor scars much transversely elongated, and an entire absence of the radiating posterior costæ. The general characters are so much alike in the two, that I shall look forward with much curiosity to future descriptions of these shells.

Loc. and Hor.—Jamberoo, and Black Head, Illawarra District —Upper Marine Series.

STUTCHBURIA SIMPLEX, *Dana*, sp.

Modiolopsis simplex, Dana, Am. Journ. Sci., iv., 1847, p. 159.

Cypricardia simplex, Dana, Wilkes' U. S. Explor. Exped., x., 1849, p. 703, pl. 9, f. 2.

Obs.—Four shells in our collection correspond in outline and size with the above species of Dana's, but with the internal characters agreeing in every respect with those of *Stutchburia*, as for instance those of the hinge, adductor impressions, and palial lines. The only points of departure are the size, a more truly oblong shape, and the exterior simple, sub-plicate, and not at all radiate. In the absence of Dana's type, it is, of course, impossible to speak with certainty, but I am strongly of opinion that his species appertains to the present genus.

Loc. and Hor.—Wollongong, Illawarra District; Jervis Bay, Shoalhaven District.—Upper Marine Series.

STUTCHBURIA FARLEYENSIS, *sp. nov.*

(Pl. xxxii., figs. 3 – 6.)

Sp. Char.—Shell transversely elongated, oblong to almost quadrangular, moderately compressed, average length one and three quarter inches, breadth one inch ; dorsal and ventral margins sub-parallel, the former straight, and not quite as long as the vales, the latter slightly insinuated near the middle, and expanding posteriorly; anterior ends very small, margins slightly oblique from the umbones downwards, but in some examples almost straight walled ; posterior ends much compressed, margins well and gently rounded ; valves most convex about widway between the umbones and posterior termination of the hinge lines; posterior ridges very obtuse, dying out on the compressed posterior ends, above and between them and the dorsal margins the surface of the valves is somewhat hollowed, and before them are shallow ill-defined cinctures dying off towards the insinuated points on the ventral margins; umbones inconspicuous; anterior adductor impressions triangular, of medium size but strongly marked, deep anteriorly and superiorly, with well marked bounding grooves on the posterior sides; posterior adductor impressions inconspicuous, flattened, placed high up under the hinge lines, and immediately at the ends of the dorsal margins : indications of scars exist within the umbonal cavities. Ligamental fulcral impressions wide and shallow; ligamental pits transversely elongated, each giving off an oblique and posteriorly directed ridge, and forming the anterior boundaries of the shallow cinctures. Pallial scars well marked, continuous (*i.e.*, not broken up), the surfaces below rapidly thining away o the ventral margins.

Obs.—All the specimens are in the form of internal casts, as an impure somewhat concretionary limonite, allowance must therefore be made in applying the above description to future examples with the test preserved. What the nature of this envelope was we are ignorant, but on a few of the specimens there are apparently faint indications of posterior radiating costæ. An example from the Upper Marine Series of Wollongong, possessing the outline and measurements of this species, and with the test preserved, exhibits a few radiating posterior costæ and strong imbricating laminæ of growth that may represent the more perfect condition of *S. farleyensis*, but it cannot be accepted as by any means certain.

Dana described two shells as *Cardinia ? recta* and *C.? cuneata,*[*] both from the Illawarra District differing greatly in outline from those forms I have made typical of the new name *Stutchburia*, but the internal features depicted in his figures are precisely similar to those of *S. farleyensis*. They seem to be edentulous, and the only point allying them with *Cardinia* are the minute posterior ends. It is possible, therefore, that the shells in question may be species of *Stutchburia*, in which case the generic characters of the latter, will of necessity require to be slightly modified.

The internal casts of *S. farleyensis* occur in great numbers in the Lower Marine Series at Farley, near West Maitland, and it is essentially a Lower Marine species, but the Geological Survey Collection contains a shell from the Upper Marine Series of Richmond Vale, Parish of Stanford, County Northumberland, of somewhat larger dimensions than the measurements above given; otherwise it agrees in every detail with my description. This bears out the suggestion that the shell found at Wollongong, with the test preserved is also *S. farleyensis*.

---

* Dana—Wilkes' U.S. Explor. Exped., x., 1849, pl. 4, f. 5a, b, and f. 6a – e.

Loc. and Hor.—Railway Cutting at Farley, near West Maitland —Lower Marine Series; ? Wollongong, Illawarra District, and ? Richmond Vale, as above—Upper Marine Series.

STUTCHBURIA OBLIQUA, *sp. nov.*

(Pl. xxxi., fig. 3.)

Sp. Char.—Shell transversely obliquely oblong, slightly modioliform ; length two to two and a quarter inches, depth one and one-eighth to one and a quarter inches ; valves moderately convex, narrowing anteriorly, and expanding to some extent posteriorly; dorsal and ventral margins sub-parallel, the former straight, but not as long as the shell, the latter with slight inflections anterior to the greatest convexity of the valves ; anterior ends remarkably small, the margins obliquely rounded, posterior ends becoming flattened, the margins obliquely rounded above and below; greatest convexity anterior to the valve centres, with ill-defined cinctures from the umbones, which are almost terminal ; ligamental fulcral grooves well marked ; anterior adductor scars small, somewhat triangular and immediately beneath the umbones, with slightly thickened posterior margins, posterior adductor scars inconspicuous; sculpture consisting of well marked close concentric laminæ, arranged in broad growth zones, crossed by radiating costæ (six in one example, ten in another), all posterior to the shallow cinctures, and widening from one another on and above the diagonal ridges, with a generally roughened surface.

Obs.—This species differs from all the foregoing forms in its obliquity, and somewhat modioliform outline. It resembles *S. costata* in the presence of the posterior radiating costæ, but the two cannot otherwise be mistaken for one another. It is a comparatively much broader species than either *S. simplex* or *S. farleyensis.* It is known to me both in the testiferous condition, and as an internal cast, the former being in the collection of the Geological Survey, the latter in our own.

Loc. and Hor.—Jervis Bay, Shoalhaven District (cast)— Upper Marine Series; Farley (testiferous)—Lower Marine Series.

Genus PLEUROPHORUS, *King*, 1844.

(Ann. Mag. Nat. Hist., (1), xiv., 1844, p. 313,)

Obs.—Notwithstanding the fact that the shells referred to this genus by De Koninck and myself do not fall within its limits, we still have, I believe, a true and undescribed *Pleurophorus* in our Permo-Carboniferous rocks. It occurs commonly at Farley with *Stutchburia farleyensis,* and is often mistaken for it, a little examination, however, will at once enable the difference between the two to be detected.

In *Pleurophorus*, as described by King, and redescribed by Waagen,[*] the equivalve closed shell possesses two cardinal inter-locking teeth in each valve, and a posterior lateral one, extending the entire length of the hinge; there is a lunule and an escutcheon, entire pallial lines, and fairly well marked adductor impressions, the anteriors having before them strong shelly ridges.

Waagen has pointed out that "one of the two cardinal teeth is often very little developed," and such is the case in most of our specimens, but in a cast in the Geological Survey Collection, the impressions of all four teeth are distinctly visible.

PLEUROPHORUS GREGARIUS, *sp. nov.*

(Pl. xxxiii., figs. 2 – 5.)

Sp. Char.—Shell transversely elongated, oblong, robust, practi-cally maintaining the same width throughout its whole length, the latter on an average one and three quarter inches, depth one inch; dorsal and ventral margins straight, parallel; bodies of the valves convex, most so at about the middle, but the flanks rather flattened or straight walled: faint cinctures exist, cutting the ventral margins at about the centre; anterior ends small, the margins convexly rounded; posterior ends but slightly flattened, the margins rounded; umbones conspicuous and incurved, a little flattened above; escutcheon long, widening posteriorly; lunule apparently cordiform, shallow; posterior cardinal teeth below the umbones, the most anterior of the left valve often inconspicuous; posterior lateral teeth leaving deep impressions in casts, the left often double; anterior adductor impressions deep, low in position, forming strong prominences in the cast, guarded by a posterior shelly ridge, which varies in intensity in individuals; posterior adductor impressions faintly marked, continuous; exterio-pallial margins flattened, leaving very conspicuous impressions in casts; sculpture of concentric laminae, no radii.

Obs.—With one exception this is only known to me as casts, and in the adult state I find the measurements very constant. The exception referred to, otherwise possessing all the characters of the species, is two and three quarter inches long by one and a half deep. I have only seen one individual that may be *P. gregarius* with the test preserved, but it is from a different horizon. The sculpture is concentric, with well marked laminae, but without any traces of radiating costae. *P. gregarius* belongs to the group Imbricati in the classification of Waagen;[†] and in outward form resembles to some extent all three species placed by him therein, but is a broader and more robust form.

* Waagen—Mem. Geol. Surv. India, Pal. Ind. (13), iii., 1881 (Salt Range Fossils, Pelecypoda) p. 214.
† Waagen, *loc. cit.*, p. 216.

Loc. and Horizon.—Farley, near West Maitland—Lower Marine Series ; ? Wollongong, Illawarra District—Upper Marine Series.

Genus LIMOPTERA, *J. Hall.*

(35th Ann. Report N.York State Mus. Nat. Hist., 1884, p. 406*a*.)*

LIMOPTERA ? PERMOCARBONIFERA, *sp. nov.*

(Pl. xxxii., figs. 1 - 2).

Sp. Char.—Shell obliquely subrhomboidal, length and width almost equal, but the latter somewhat the greater, produced postero-ventrally; valves very unequal, the left convex, the right more or less flattened, but the greatest convexity of the latter immediately below the umbone; hinge line straight, probably as wide as the shell ; ligamental area not well preserved, but apparently wide and deep beneath the umbones, and narrow posteriorly; anterior ends or auricles flattened in both valves, separated from the bodies of the valves by sharp declivities, the anterior margins below obliquely and sharply rounded ; posterior ends or wings triangular, flattened, much larger in the left than the right valve, distinctly demarcated from the bodies of the valves, margins sharply emarginate, then swelling out to round the protuberant postero-ventral portions. Left umbo prominent, nearly central in position, the umbonal region abrupt on the anterior, but gently sloping on the posterior side to form a posterior slope ; umbonal cavity of the right valve containing a number of nodes (in the cast) indicating pits for muscular attachment ; adductor impressions and pallial scars not distinctly marked ; sculpture of the left valve consists of irregular concentric laminæ and faint oblique radii, extending from the umbonal centre well on to the posterior end ; the surface of the right valve is transversely wrinkled on the cast.

Obs.—The specimen is somewhat mutilated, but it presents most of the principal characters of the genus *Limoptera*, with the exception of the cardinal folds and the oblique posterior tooth. The former however may be hidden by the matrix infilling the deep ligamental recess beneath the umbones. The precise generic affinity of this shell, I am not at present prepared to give, but it accords better with Hall's definition of *Limoptera* than with any other similar genus. It is more produced posteriorly than any of the shells figured by Hall under this name, and is also specifically distinct from any other yet described from New South Wales. The outward form only is that of some *Glyptodesmœ*, or *Pterinea* as restricted, or even more so perhaps *Leiopteria* or *Leptodesma*,

---

* It is impossible to unravel the mystery surrounding the first annunciation of many of the late Prof. James Hall's genera. This reference is simply given as one to a description of the genus.

particularly in the case of the last with its deeply emarginate posterior wing.

Loc. and Horizon.—Mouth of Crooked River, near Gerringong, Illawarra District—Upper Marine Series.

Genus MYTILOPS, *J. Hall.*

(1st Report State Geol. N. York, 1884, p. 15.)*

MYTILOPS ? RAVENSFIELDENSIS, *sp. nov.*

(Pl. xxxiii., figs. 6 – 7).

Sp.Char.—Shell (left valve) narrow, somewhat elliptical, oblique, generally mytiliform, gibbous and transversely arched posterior to the umbone; hinge line faintly arched in the cast, but less than the length of the shell; no ligamental furrows, but beneath the umbo is a single oblique cardinal fold, posterior teeth or folds none; ventral margin very obliquely inclined; umbone terminal, no anterior end; posterior end, or general body of the shell convex immediately behind the umbone, gradually flattening towards the rounded posterior margin, where the valve is broad; anterior muscular impression, invisible and the posterior very faint and high; pallial impression well marked and continuous, the exterio-pallial margin wide; sculpture unknown, but probably concentric and non-radiate.

Obs.—Allied either to *Mytilops* or *Mytilarca*. It possesses the outline of the latter genus, and approaches the former in its simpler hinge structure; differs from *Mytilarca* in the absence of a ligamental area and posterior teeth, and from *Mytilops* in the presence of the cardinal fold. So far as our Permo-Carboniferous fauna is concerned this is again an undescribed form.

Loc. and Horizon.—Ravensfield Quarry, near Farley—Lower Marine Series.

---

* See note to preceding page.

RECURRENCE OF *MEGADERMA GIGAS*, DOBSON.

By EDGAR R. WAITE, F.L.S., Zoologist.

Twenty years ago Dobson described *Megaderma gigas*, a new species of Bat from Australia,[*] since which time no further example has been made known. The type taken at Mount Margaret, Wilson's River, Queensland, is a male and is in the Göttingen Museum.

On the 9th February, the Trustees received from Mr. Fred. Hogan, by presentation, a specimen of the same species, taken in the Pilbarra District, Northern West Australia. This is a female, and presents some few differences from the description of the male. The example was mounted before reaching us, so that in the following table of dimensions, the measurements of the length of the head and body are approximate only; the other dimensions are, however, absolute.

Dobson's measurements are recorded in inches and tenths—these I have reduced to millimetres for comparison with my own figures, so expressed. From these it will be seen that the female, which is adult, is generally smaller than the male, but the lengths of the tibia and the first phalanx of the fifth finger are actually greater; more striking, perhaps, is the relative difference in the phalanx of the second finger, but this supports and emphasises Dobson's statement— "While in *M. spasma* the extremity of the second finger does not extend as far as the middle of the first phalanx of the third finger, in this species [*M. gigas*] as in *M. frons*, it extends beyond it."

Further evidence that the West Australian example is referable to *M. gigas* is supplied by the circumstance of the extremity of the carpus, the thumb, and the membrane between the thumb and the second finger being hairy, in which respect it differs from the other known species.

The mammæ are two in number; they are situated one on each side of the upper abdominal region.

The colour does not differ from Dobson's desciption, but the pale grey of the upper surface shows brownish tints in certain lights; there is now no indication of the deep blood-red colour at the anterior base of the ears, shown in Dobson's figure and described as being present in the type when obtained, but which had apparently faded out before the author saw the specimen.

---

[*] Dobson—Proc. Zool. Soc., 1880, p. 461, pl. xlvi.

The general colour of the specimen, were it not characteristic of the genus, would have suggested albinism. The effect of the beautiful whiteness of the fur of the head and whole under surface has been quite lost in the figure.

Dimensions.

	♂ Type.	♀ Aust. Mus.
Head and body	... 135·0 mm.	110·0 mm.*
Length of head	48·5 ,,	41·0(?),, †
Ear	56·0 ,,	47·0 ,,
,, tragus, anterior lobe	12·0 ,,	10·0 ,,
,, ,, posterior lobe	25·5 ,,	22·0 ,,
Nose leaf	16·0 ,,	15·0 ,,
Forearm	... 117·0 ,,	103·5 ,,
Thumb, or first finger ...	21·0 ,,	19·5 ,,
Second finger, metacarpal	84·5 ,,	80·5 ,,
,, ,, phalanx ..	16·0 ,,	17·0 ,,
Third finger, metacarpal	69·0 ,,	68·0 ,,
,, ,, 1st phalanx	47·0 ,,	44·0 ,,
,, ,, 2nd ,,	92·0 ,,	80·0 ,,
Fourth finger, metacarpal	79·0 ,,	78·0 ,,
,, ,, 1st phalanx ...	25·5 ,,	24·5 ,,
,, ,, 2nd ,,	38·5 ,,	33·0 ,,
Fifth finger, metacarpal	84·5 ,,	82·0 ,,
,, ,, 1st phalanx	32·0 ,,	33·0 ,,
,, ,, 2nd ,,	28·5 ,,	23·5 ,,
Tibia	44·0 ,,	45·0 ,,
Calcaneum	28·5 ,,	28·0 ,,
Foot... ...	28·5 ,,	23·5 ,, ‡

* From crown of head.
† Occipital region of skull removed.
‡ Exclusive of claw.

An EXTENDED DESCRIPTION of *MUS FUSCIPES*, WATERHOUSE.

By Edgar R. Waite, F.L.S., Zoologist.

(Figs. 1 – 4).

Few of our native rats have been described, other than from external characters, and such characters are in many cases of but secondary value. As a revision of the Australian Muridæ is much needed, any effort towards the completion of specific descriptions will be welcomed by the Monographer.

By the kindness of Mr. E. G. W. Palmer, we are able to supply deficiencies in our knowledge of *Mus fuscipes*. The specimens described were taken at Lawson, on the Blue Mountains, and of them my correspondent writes :—

"So far as my observations go, they are locally rare, but there is a small colony in my orchard, which I first observed about twelve years ago. Dogs and Dasyures have checked their rapid increase. A peaty ridge is their favorite burrowing place, and they burrow to a great depth. They make long well-cleared surface runs, so that their burrows are easily found. Water seems very necessary to them, and they swim freely. They feed on grasses and herbage, and consume or injure much fruit, climbing the trees for it or nibbling the windfalls, which they carry to the drains and watercourses. From dissections, I believe they seldom have more than two or three young at a time. Their teeth are very powerful, and they make good use of them when roots or dead timber obstruct their excavating. Just now (August 16th, 1899), they seem to be hibernating, as they rarely come out of their nests."

Subsequently Mr. Palmer told me that the rats had left their old haunts, or more probably had been cleared out by snakes, as a large Black Snake (*Pseudechis porphyriacus*) had been frequently seen in the immediate neighbourhood. It had, so far, evaded capture. At a recent meeting of the Linnean Society of New South Wales, Mr. Palmer announced that he had been bitten by a Black Snake in his grounds at Lawson.*

Description.—Fur long, very thick and soft to the touch. Colour rather variable, from yellowish-brown to blackish-brown. Basal portion of the fur deep grey, almost black, the tips yellow, sometimes

---

* Abstract Proc. Linn. Soc. N.S.W., 28 Mar., 1900.

inclining to reddish, thickly interspersed with long black hairs. Muzzle and face grey, whiskers dark brown or black. On the sides of the body the black hairs are less numerous and shorter, and the yellow colour lighter. On the lower surface the basal fur is grey, not so dark as on the back, and more broadly tipped with yellow, which however is very pale. Ears small, almost naked without, and sparingly clothed with short yellowish-brown hairs within; laid forward, they fail to reach the eye by their own length. Limbs thickly clothed with greyish-brown hairs. Tail shorter than the head and body; the hairs are longer than two scales, but do not conceal them, they are not particularly stiff and are of black colour; twelve scales to the centimetre. Mammæ $2 + 3 = 10$.

Dimensions.

	A. ♂	B. ♀
Head and body	173·0 mm.	176·0 mm.
Tail	116·0 ,,	119·0 ,,
Length of head	45·5 ,,	44·0 ,,
Muzzle to ear	36·6 ,,	34·0 ,,
Ear	21·0 ,,	20·0 ,,
Forearm and hand	39·5 ,,	41·5 ,,
Hind foot	31·2 ,,	29·3 ,,
Heel to front of last foot pad	15·0 ,,	14·5 ,,
Last foot-pad	4·0 ,,	3·6 ,,

Skull.—Stout, compared with *Mus decumanus*, deeper and considerably more arched, the nasal region shorter and thicker; the nasals do not project beyond the line of the premaxillary, they are wide in front but taper backwards, so that the posterior is only one-third the width of the anterior portion. The narrowed part is sunk below the level of the premaxillaries which form a ridge on each side. The supraorbital

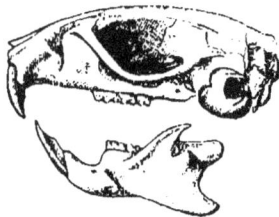

Fig. 1.

ridge is very marked, and forms a distinct beading, but loses that character on the temporal region. The interparietal is short, but of average width, its front margin forming a nearly straight suture with the parietals. The anterior palatine foramina are narrow and extend backwards to the anterior margin of the first molars. The anterior zygoma root has the angle rounded, the front edge is vertical and slightly concave; the foramen magnum is wider than deep. The mandible is very powerful, with strong muscle ridges; in front of the anterior molar it is nearly vertical, and the incisor capsule is large and deflected outwards.

Teeth.—The front edges of the upper incisors are of particularly deep orange colour, the lower ones are somewhat paler. The molars are relatively and actually larger than those of *Mus decumanus*, and are noticeably broader ; the upper series converge anteriorly, and are somewhat bowed outwards.

Fig. 2.   Fig. 3.   Fig. 4.

Dimensions of Skull.

	A. ♂	B. ♀
Greatest length	40·3 mm.	39·6 mm.
Basal length	36·6 ,,	36·2 ,,
Greatest breadth	21·9 ,,	21·7 ,,
Nasals, length...	15·0 ,,	14·4 ,,
,, greatest breadth... ...	4·6 ,,	4·6 ,,
Interorbital breadth	5·3 ,,	5·6 ,,
Interparietal length	4·1 ,,	4·6 ,,
,, breadth	10·3 ,,	10·6 ,,
Brain case, breadth	17·2 ,,	17·3 ,,
Anterior zygoma root	5·3 ,,	5·5 ,,
Diastema...	10·4 ,,	10·7 ,,
Palate, length...	21·9 ,,	21·7 ,,
Anterior palatina foramina ...	7·2 ,,	7·1 ,,
Upper molars, length	8·1 ,,	8·3 ,,
Lower ,, ,, ...	7·6 ,,	8·0 ,,
Condyle to incisor tip	28·2 ,,	28·2 ,,
Coronoid tip to angle	12·2 ,,	11·4 ,,

The caudal vertebræ are twenty-five in number ; in the longer tailed *Mus arboricola* (*M. rattus, fide* Thomas), the vertebræ number thirty-eight.

Waterhouse described the colour of the lower incisors as black, evidently a peculiarity of the individual examined. Gray, writing on *Mus lutreola* remarks, "front teeth yellow"; while Gould says, "the incisor teeth are orange-coloured." I do not remember having examined a rat's skull in which the incisor teeth are so deeply tinted. The writers quoted describe the under parts of the body as being greyish-white, or grey lead colour. In all the fresh examples I have seen, yellow enters noticeably into the colour of the ventral fur, and the almost blue colour of Water-house's figure is certainly never seen in this species. The dark frontal streak of Gould's drawing is intended to illustrate the convergence of the hairs to the centre of the head, but there is no colour band there as might be inferred.

CORRECTION.

Page 193.

For Plates xxxv.–xxxvii.

Read Plates xxxiv.–xxxvi.

forms."

It was my intention to prepare a complete list of the known Fish-fauna of the island, but I notice that Mr. Ogilby, in 1898, also in the paper quoted, writes—"As it is, the list as it now stands needs careful revision, but I hope within the next few months to be in a position to lay before the Society a thoroughly revised and enlarged catalogue of the fish fauna of the island." Under these circumstances I will leave the field open to Mr. Ogilby, and publish the following list of additions in order that his "revised and enlarged catalogue" may include the Museum records, not otherwise available to him.

Washed by a warm southerly current, Lord Howe Island supports a much more tropical fauna than is met with in lower latitudes on the mainland. It lies in latitude S. 31° 33′, and on the west side possesses an extensive coral reef. On the mainland no coral reef is found south of Stradbroke Island in Queensland,

---

* Ogilby—Aust. Mus. Mem., ii., 1889, Fishes, pp. 52–74.
† Ogilby—Proc. Linn. Soc. N.S.W., xxiii., 1898, p. 731.

Teeth.—The front edges of the upper incisors are of particularly deep orange colour, the lower ones are somewhat paler. The molars are relatively and actually larger than those of *Mus decumanus*, and are noticeably broader; the upper series converge anteriorly, and are somewhat bowed outwards.

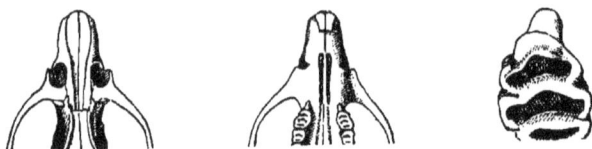

„ breadth	10·3	„	10·6	„
Brain case, breadth	17·2	„	17·3	„
Anterior zygoma root	5·3	„	5·5	„
Diastema...	10·4	,,	10·7	„
Palate, length...	21·9	„	21·7	„
Anterior palatina foramina ...	7·2	„	7·1	„
Upper molars, length	8·1	„	8·3	„
Lower „ „	7·6	„	8·0	„
Condyle to incisor tip	28·2	„	28·2	„
Coronoid tip to angle	12·2	„	11·4	„

The caudal vertebræ are twenty-five in number; in the longer tailed *Mus arboricola* (*M. rattus*, *fide* Thomas), the vertebræ number thirty-eight.

Waterhouse described the colour of the lower incisors as black, evidently a peculiarity of the individual examined. Gray, writing on *Mus lutreola* remarks, "front teeth yellow"; while Gould says, "the incisor teeth are orange-coloured." I do not remember having examined a rat's skull in which the incisor teeth are so deeply tinted. The writers quoted describe the under parts of the body as being greyish-white, or grey lead colour. In all the fresh examples I have seen, yellow enters noticeably into the colour of the ventral fur, and the almost blue colour of Waterhouse's figure is certainly never seen in this species. The dark frontal streak of Gould's drawing is intended to illustrate the convergence of the hairs to the centre of the head, but there is no colour band there as might be inferred.

ADDITIONS TO THE FISH-FAUNA OF LORD HOWE ISLAND.

By EDGAR R. WAITE, F.L.S., Zoologist.

(Plates xxxv. - xxxvii. ; and Figs. 1, 2).

Since 1889, when the Fish-fauna of Lord Howe Island was first published in collected form,[*] sundry additions have been recorded by Mr. J. D. Ogilby and myself. The former in his latest contribution writes[†]—"The present additions bring the number of species recorded as inhabiting or visiting the shores of the island up to one hundred and thirteen, with seven (or six) unidentified forms."

It was my intention to prepare a complete list of the known Fish-fauna of the island, but I notice that Mr. Ogilby, in 1898, also in the paper quoted, writes—"As it is, the list as it now stands needs careful revision, but I hope within the next few months to be in a position to lay before the Society a thoroughly revised and enlarged catalogue of the fish fauna of the island." Under these circumstances I will leave the field open to Mr. Ogilby, and publish the following list of additions in order that his "revised and enlarged catalogue" may include the Museum records, not otherwise available to him.

Washed by a warm southerly current, Lord Howe Island supports a much more tropical fauna than is met with in lower latitudes on the mainland. It lies in latitude S. 31° 33′, and on the west side possesses an extensive coral reef. On the mainland no coral reef is found south of Stradbroke Island in Queensland,

---

* Ogilby—Aust. Mus. Mem., ii., 1889, Fishes, pp. 52 – 74.
† Ogilby—Proc. Linn. Soc. N.S.W., xxiii., 1898, p. 731.

latitude S. 27° 25', but the fauna there is in no way comparable with that of Lord Howe Island, which is endowed with many purely tropical forms.

The temperate fishes are, in the main, those found on the coast of New South Wales, while the tropical ones generally suggest a Melanesian rather than an Australian origin.

Several bathy-pelagic fishes have already been recorded from the island; we may mention *Isistius*, *Brama*, *Gempylus*, and *Tetragonurus*. Being dependent for novelties on residents or visitors to the island, who, though kind and well meaning, can never accomplish the results of a trained collector, our work of building up a knowledge of the fauna is necessarily slow.

If a small steamer, fitted up for dredging, could be chartered for a few weeks, we should, I am confident, be able to add many new and interesting forms both among fishes and invertebrates.

The present contribution is based on material received at the Museum by the kind offices of Mrs. T. Nichols and her daughters, Mr. T. R. Icely, J.P., late Visiting Magistrate, Messrs. Wm. Nichols, and J. B. Waterhouse, residents on the island, and the late Mr. W. E. Langley.

The paper deals only with unrecorded species, two excepted, namely:—*Trachinotus russelli*, Cuvier and Valenciennes, and *Chironemus marmoratus*, Günther, previously incidentically mentioned in a publication not generally distributed.

Including an unnamed *Atopichthys*, thirty-two species are recorded, of which four are described as new, namely:—

> *Amphiprion latezonatus.*
> *Holacanthus conspicillatus.*
>   ,,   *semicinctus.*
> *Euchilomycterus quadradicatus.*

The last-named is regarded as the type of a new genus. A new generic name, *Acanthocaulus*, is also proposed to replace *Prionurus*, Lacépède, pre-occupied. Of the remainder, the following are new to the Australian fauna; some have, however, been taken in Torres Straits:—

> *Leptocephalus cinereus*, Rüppell.
> *Aulostomus chinensis*, Linnæus.
> *Gempylus serpens*, Cuvier and Valenciennes.
> *Decapterus sanctæ-helenæ*, Cuvier and Valenciennes.
> *Epinephelus fasciatus*, Forskal.
> *Thalassoma aneitense*, Günther.
> *Platyglossus opercularis*, Günther.
> *Holacanthus tibicen*, Cuvier and Valenciennes.
> *Naseus unicornis*, Forskal.
> *Alutera monoceros*, Osbeck.
> *Ovoides meleagris*, Lacépède.
> *Parapercis cylindrica*, Bloch.

Those below-named, previously known from Australia, are new to the fauna of New South Wales, of which colony Lord Howe Island is a dependency:—

Isistius brasiliensis, Quoy and Gaimard.
Hippocampus hippocampus, Linnæus.
Epinephelus tauvina, Forskal.
Therapon jarbua, Forskal.
Ostracion cubicus, Linnæus.

[SISTIUS BRASILIENSIS, *Quoy and Gaimard.*

(Figs. 1, 2).

Under the name *Leius ferox*, Kner[*] well described this species from Australia, but without any more definite habitat. It has not been since recorded from our waters. The example now under notice was sent to the Trustees from Lord Howe Island by Mrs. T. Nicholls, and forms a most interesting addition to the fauna of the Island. In his Whaling Voyage, F. D. Bennett[†] described, as *Squalus fulgens*, two examples taken at different periods of the voyage by means of a tow net, proving the pelagic habit. The largest of these was an adult female, and measured eighteen inches in length. Ours is a male, and measures 390 mm. (= $15\frac{1}{4}$ inches), it possesses the dark band across the chest, and the white edged fins of *Scymnus torquatus*, Valenciennes.

The upper teeth are arranged in thirty-three rows, in a band of crescent shape, four or five deep mesially, and two or three laterally; each tooth is strongly curved outwards and backwards, and the whole series is depressible. The palate is very hard and is evidently the counterpart of the tongue which is furnished with a similarly hard plate extending all along the front and lateral margins. Such crushing surfaces would appear to be unnecessarily developed if used only for reducing the weak shells of *Ianthina* and *Reclusia*, or even the armament of *Nautilograpsus*.

The lower teeth form a single functional fixed series arranged in rather more than a semicircle, with a diameter of 35·5 mm. in the specimen examined. Each tooth consists of a thin erect plate with a triangular apex, the margins of which are smooth, the basal portion is faintly striated and has a central pit connected with the basal edge by a short channel ; the median tooth is wholly exposed and bi-symmetrical (fig. 1), the lateral teeth are imbricate and their apices directed away from the symphysis, on approaching the angle of the mouth they become smaller. There are fifteen teeth on each side of the median one, making thirty-one in all.

Fig. 1.

* Kner—Denks. Akad. Wiss. Wien. xxiv., 1865, p. 10, pl. iv., fig. 2.
† Bennett—Whaling Voyage, ii., 1840, p. 255.

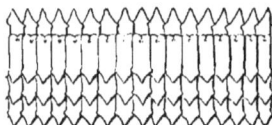

Fig. 2.

On the inner side of the jaw there are three other rows of precisely similar teeth, apparently destined to successively replace those in use, but not functional until the existing row has been lost. When the second row of teeth has reached the summit of the alveolus, erection must be speedy, for the dentary is very thin, nowhere more than 2·5 mm. in thickness. The teeth in reserve have their apices pointing directly downwards, and the bases of the first reserve row are applied a little above the centre of the teeth of the functional row (fig. 2).

As this paper was passing through the press the Trustees received the Report on the Deep Sea Fishes of the "Albatross"; this splendid work contains a descriptive notice and figures of *Isistius*.* Garman there enumerates all the examples known, from which it would appear that ours ranks as the fourteenth. The "Albatross" specimen was taken at Station 3413, where a depth of 1,360 fathoms was registered. Some idea is however expressed that the fish may have been netted during the ascent of the trawl, at a less depth.

It is of interest to notice that Garman's suggestion that the number of teeth may within certain limits increase with age, receives support from the characters of our specimen. Examples previously taken were found to have twenty-six or fewer teeth in the lower jaw, such were however, immature, being ten inches or less in length. In the individual taken by the "Albatross," which measures more than eighteen inches in length, there are thirty-one such teeth, precisely the number possessed by the Lord Howe Island specimen, as before stated. This, though not quite so large is possibly adult, as indicated by the nature of the sexual organs. The largest example recorded measures more than nineteen and a half inches in length, while Garman remarks—"The species is mature at a length of eighteen inches."

CONGERMURÆNA HABENATA, *Richardson.*

An example obtained by Mr. T. R. Icely quite agrees with specimens from the mainland, having the tail proportionately longer than in Richardson's type from New Zealand. This difference has been expressed by Ramsay and Ogilby in the name *C. longicaudata.*

---

* Garman—Mem. Mus. Comp. Zool., xxiv., 1899, p. 34, pls. i., ii., iii., lxix.

LEPTOCEPHALUS CINEREUS, *Rüppell.*

Conger marginatus, Valenciennes.

This species has been recorded from Torres Straits. The island example differs only from the descriptions in being of less uniform colouration ; the ground colour is grey, crossed by about fifteen irregular more or less complete bands of a darker tint. There is a jet black spot near the tip of the pectoral, on the inner side of which it becomes a much more extensive blotch. The specimen, preserved in formol, has the colours unusually well retained.

ATOPICHTHYS, *Garman.*

"Heretofore certain pelagic, much compressed, band-like, translucent to transparent, larval fishes, have been placed in the genus Leptocephalus of Gronow, 1763. The type of the genus is Leptocephalus Morrisii, Penn., 1776, a larval form which has lately been traced to its adult in Murœna conger, Linn., 1758, which again was the typical species of Risso's genus Conger, 1826. In consequence Leptocephalus has taken the place of Conger as the title of the genus, and many of the Leptocephalids which do not belong to that genus, and cannot yet be definitely located, are left unnamed. That there is a considerable number of these larval forms that cannot be placed in Leptocephalus, but that belong to various other genera not now determined with sufficient accuracy, is evident enough from the figures and descriptions given below. Rather than to assign them at random, it is here proposed to form a group for these and similar unplaced larvæ, Atopichthys, in which they may remain until such time as by means of larger collections the adult forms and their respective generic affinities may be determined."[*]

As this is perhaps the first occasion on which the name *Atopichthys* has been used since characterised, I have reprinted Garman's remarks in full.

In January last the Trustees received from the island a very fine larva, collected by Miss Nicholls. I do not propose to name this form, and will merely indicate some of its leading features :—

Body elongate, narrow and slender ; greatest depth behind the middle, one-fifteenth of the entire length. Head more than twice as long as high, 13·3 in the total length. Snout pointed, nearly one-fourth of the head. Eyes lateral, 7·5 in the head, situated in its anterior half. Mouth large, reaching to below the middle of the eye. Upper jaw the longer. Teeth small, inclined backwards. Gill-opening narrower than the eye, extending below the base of the pectoral. About one hundred and fifty muscle bands.

---

[*] Garman—*Loc. cit.,* p. 325.

Pectoral broad, longer than the eye: no filamentary caudal. Dorsal and anal fins indistinct, better defined posteriorly. Translucent; small black spots form a band on the nape and a similar one on the throat.

Total length	... ...	200 mm.
Height of body...	...	13 ,,
Length of head...	...	15 ,,
,, snout	...	4 ,,

AULOSTOMUS CHINENSIS, *Linnæus.*

Mr. T. R. Icely obtained a nice example in December, 1892. It measures 560 mm. in length, and has the following characters :—

D. xii. 25. A. 25. P. 16. V. 6. C. 15 + 2.

Length of head 3·2, height of body 11·0 in the total length, exclusive of caudal. Eye 3·0 in the postorbital part of the head. Lower jaw prominent, with the barbel one-half longer than the diameter of the eye. Premaxillary slender ; maxillary narrow anteriorly, greatly broadened behind, its posterior margin notched. Upper jaw edentulus ; lower jaw with a small patch of minute teeth in each ramus. Caudal pedicle equal in length to the distance of the posterior margin of the opercle from the centre of the eye. Ventrals short, equal to the least depth of the snout; they extend to the vent, which is situated midway between the hinder edge of the opercle and the base of the caudal rays.

When freshly obtained, the colour was pink and the fins yellow. The body is longitudinally streaked, the streaks disposed both above and below the lateral line ; there is a deep black bar across the centre of the maxillary, and a narrower one passing through the nostrils to the eye ; a black spot at the base of each ventral fin, and another on the upper caudal rays. The bases of the dorsal and anal fins are black, and this colour is continued up the front margin of the dorsal; the portion of the body between these fins is very dark, relieved by two of the white body streaks in a line with the upper and lower margins of the caudal pedicel, each streak with two ganglion-like spots; two similar but fainter spots exist at the bases of the fins. There are also three pairs of spots on the pedicel, forming transverse bands.

Drs. Jordan and Evermann, in describing the family Aulostomidæ, write[*]:— "A single genus, with two species, found in tropical seas." This should surely read "three species," for they mention *A. maculatus* and *A. cinereus*, neither of which is synonymous with *A. chinensis*, admitted as the type of the genus.

---

[*] Jordan and Evermann—Bull. U.S. Nat. Mus., 47, 1896, p. 754.

MACRORHAMPHOSUS GRACILIS, *Houttyn.*

The opinion expressed by me that this species may be of more pelagic habit than *M. scolopax*,* receives some support from the fact that an example taken on the beach at Lord Howe Island by Mr. Wm. Nichols proves to be *M. gracilis.*

HIPPOCAMPUS HIPPOCAMPUS, *Linnæus.*

A specimen from the island, registered under the synonym *H. antiquorum*, Leach, cannot, so far as I can see, be distinquished from this European species. It has been recorded from Cape York.

GEMPYLUS SERPENS, *Cuvier and Valenciennes.*

The example now recorded was obtained per Mr. Ieely in May 1893, and measures 670 mm. in length. It differs in no way from specimens recorded from the Atlantic. Drs. Jordan and Evermann† describe the lateral line as being single, in our example it is certainly double, the lower line is raised above the pectoral and otherwise runs straight along the body as described ; the upper one arises at the same point whence the lower one starts, namely just behind the first spine, and is continued along the dorsal profile close to the fin as far as its spinous termination. This condition was described by Cuvier and Valenciennes‡ in the following words— "Sa ligne latérale est droite, continue et sans inflexions ; il y en a comme une seconde le long de la base de la première dorsale." These authors also describe the palatines as being edentulous, and I fail to find the slightest trace of palatine teeth in our specimen; the American authors on the other hand write:—"palatines with a row of small teeth."

The Lord Howe Island specimen is more nearly allied to the type of *G. serpens* the Atlantic, than to *G. coluber* the Pacific form, a circumstance which favours the view that both are referable to the same species. It is to be inferred that the type of *G. coluber* has but one lateral line, a condition found in, presumably, Atlantic specimens by Jordan and Evermann.

In the Pacific, *Gempylus* has been found near the Society and Hawaiian Islands, its distribution is now therefore greatly extended westwards. "It is generally believed to be an inhabitant of great depths," in this connection I am sorry not to be able to throw more light on the subject. The Lord Howe Island example was certainly not taken in deep water, but I am not aware whether it was caught on the line or thrown upon the beach. Of the wide

---

* Waite—Aust Mus. Mem. iv., 1, 1899, p. 60.
† Jordan and Evermann—Bull. U.S. Nat. Mus., 47, 1896, p. 884.
‡ Cuvier and Valenciennes—Hist. Nat. Poiss., viii., 1831, p. 210.

distribution of the species there can be no doubt, and it is interesting to notice that the forms having the most extensive range are either bathybial or pelagic in habit.

DECAPTERUS SANCTÆ-HELENÆ, *Cuvier and Valenciennes.*

To this species I refer a fine example obtained by Mrs. T. Nichols last year. It measures 310 mm. in length and though not in very good condition, exhibits all the features of the species with the exception of the character of the first dorsal spine, this is short, not half the length of the second and is correspondingly feeble. I have assumed that Steindachner is correct in regarding *Caranx muroadsi,* Temminck and Schlegel, as synonymous with *D. sanctæhelenæ.*

TRACHINOTUS RUSSELLI, *Cuvier and Valenciennes.*

This species is incidentally mentioned by Ogilby* as occurring at Lord Howe Island, but as the reference might easily be overlooked, attention is here drawn to it. The specimens in the Museum were received from Mr. Icely.

BRAMA RAII, *Bloch.*

Castlenau has recorded this species from Port Jackson, but doubt has since been thrown upon his identification. We have an example in the Museum, obtained from Lord Howe Island by Mr. Icely, so that Castlenau's record was doubtless correct. It has the characteristically deeply forked caudal of the species, and the radial formula is D iii. 33; A. ii. 28. Like *Lampris luna* (a notice of which recently appeared in this publication), *Brama raii* is a large pelagic fish, widely distributed and descending to considerable depths ; any new recorded habitat, though interesting, is therefore not surprising.

EPINEPHELUS FASCIATUS. *Forskal.*

This species is represented from the island by a fine example, 300 mm. in length. Previously it was known from Darnley Island in Torres Straits, which gave it a place in the Australian fauna. We have other two specimens in the collection, one of which is from Port Moresby, British New Guinea, possibly an unrecorded habitat.

THERAPON JARBUA, *Förskal.*

This widely distributed species has been recorded from the north and north-east coast of the continent, and a single example obtained by Mr. Icely, enables me to add it to the fauna of Lord Howe Island.

---

* Ogilby—Edible Fishes N.S.W., 1893, p. 90.

CHIRONEMUS MARMORATUS, *Günther*.

The Kelp-fish is common on the coast of New South Wales, and is, I learned from the Lord Howe Islanders, sometimes caught off the coral reefs. We have an example in the Museum, obtained by Mr. Icely, and this is recorded by Ogilby,[*] but for reasons applied to *Trachinotus*, is also noticed here.

AMPHIPRION LATEZONATUS, *sp. nov.*
(Plate xxxiv.)

D. xi. 15.   A. ii. 13.   V. i. 5.   P. 17.   C. 15 + 2.   L. lat. 38.
L. tr. 6/18.

Length of head, to which the caudal fin is equal, 3·6, height of body 2·0 in the length (caudal excluded). Diameter of eye 2·8, length of snout 3·5 in the length of the head. Interocular space very slightly convex, a little more than the diameter of the eye. Twelve gill rakers on the lower limb of the first arch, the centre ones narrow and rather long. Teeth conical in a single series in each jaw. Preopercle denticulated, its angle in advance of the centre of the eye. Opercle formed of two lobes strongly spinose. Dorsal fin without notch, its first spine placed above the margin of the opercle, its length less than the diameter of the eye, the eighth spine is the longest, and is one-tenth longer than the eye ; the central rays are the longest, twice the diameter of the eye. The anal spines are short but stout, and the longest rays are more posterior than the corresponding ones of the dorsal.

The pectoral and ventral are long and equal, one-seventh longer than the head, the latter extending to the base of the second anal spine. The caudal is forked, the upper rays the longer, the length of the pedicel equals its height, which is one half the length of the head.

Scales.—The scales are large, with entire margins, the lateral line, which terminates in advance of the dorsal rays is composed of thirty eight scales, this is also the number of the series between the opercle and the caudal.

Colours.—General colour dark brown with three light cross bands, the first is as wide as the diameter of the eye, the posterior margin of which it embraces ; it passes from the occiput, in front of the dorsal spines downwards and forwards, and crosses the preopercle, opercle, and sub-opercle: its posterior margin is convex and on the dorsal profile this margin is deflected forwards. The second band arises in the space between the eighth spine and the second ray, its anterior edge takes a forward sweep gaining the ventral profile just behind the ventral fins, its hinder edge is

---

[*] Ogilby—Edible Fishes N.S.W., 1893, p. 55.

C

deflected backwards to near the termination of the anal fin along
the base of which it passes to the first anal spine; this second band
is thus much wider below than above, and at its widest part equals
the distance of its anterior edge from the snout; it is not bent
backwards along any part of the soft dorsal. The third band is
across the caudal pedicel, it is wider than the first band and both
edges are concave posteriorly. All the bands are separated from
the ground colour by a narrow white line. The membranes of the
spinous dorsal are coloured according to the tint of that part of
the body whence they arise; the whole of the soft dorsal, with the
exception of the base of the first two rays, is blackish-brown edged
with yellow, the anal has no light margin ; the pectorals are
coloured like the body bands, but their bases are dusky; the
ventrals are wholly dark. The caudal rays are very deep brown,
which colour extends along the upper and lower rays, leaving a
broad lunate margin of yellow. Length of specimen 130 mm.

Some of the species of *Amphiprion* have been shown to be very
variable, it is therefore not easy to say what amount of variation
must be allowed. The example now under notice, enters section
b of Günther's synopsis* characterised by having "Three white
cross bands, the middle of which is not bent backwards above."
In colouration it differs from any previously described by having
the middle band extremely wide, and much wider below than
above, also by the margin of the brown colour of the caudal being
posteriorly concave.

THALASSOMA ANEITENSE, *Günther.*

Julis aneitensis, Günther, Brit. Mus., Cat. Fish. iv., 1862, p. 183.

Included in a small collection made by Mr. J. B. Waterhouse
is a nice example of this species, measuring 260 mm. in length.
Günther has recorded it from Norfolk Island, and it has also been
recognised from North-east Australia, so that its occurrence off
Lord Howe Island is in no way remarkable.

PLATYGLOSSUS OPERCULARIS, *Günther.*

Of this species, which appears to be nearly allied to *P. pœcilus*,
Rich., we have four examples obtained by Mrs. T. Nicholls in
August last. There are also specimens in the Museum from the
New Hebrides.

NOVACULICHTHYS JACKSONIENSIS, *Ramsay.*

The single example obtained by Mr. Icely in January, 1895,
is somewhat larger than that taken by the "Thetis" Expedition,†
measuring 210 mm. in length. As in the type, the lower

* Günther—Brit. Mus. Cat. Fish., iv., 1862, p. 3.
† Waite—Aust. Mus., Mem., iv., "Thetis" Exp., I, Fishes, 1899, p. 87,
pl. xv.

canine teeth are developed equally with the upper ones; both pairs are curved and divergent, the lower biting between the upper ones which are rather widely spaced. The second dorsal spine is united to the third, as shown in my figure.

CHELMO TRUNCATUS, *Kner*.

A beach-dried example was obtained by Mrs. T. Nicholls, and forwarded to us in August last; and in the "old collection" there is a specimen in spirits, also from the island.

HOLACANTHUS TIBICEN, *Cuvier and Valenciennes*.

All the specimens received from the island possess but three anal spines, as in normal examples, and have fourteen dorsal spines. This record is an addition to the Australian fauna, and we are indebted to Mr. Icely for the series obtained.

HOLACANTHUS CONSPICILLATUS, *sp. nov.*

(Plate xxxv.)

D. i. + xiii. 18.   A. iii. 18.   V. 1·5.   P. 17.   C. 15 + 2.

Length of head 4·4, of caudal fin 5·4, height of body 2·0, in the length (caudal excluded). Diameter of eye 3·7, length of snout 3·2, of the preopercular spine 2·0 in the length of the head. The interorbital space is convex, one-fourth more than the diameter of the eye. Thirteen gill-rakers on the lower limb of the first arch, all small, triangular in shape. The teeth are cardiform, and are arranged in a dense band across the front in each jaw. The individual tooth is brown, with a lighter apex, broader than the base; it is tricuspid, the centre cusp long and acute, the lateral ones small and slightly deflected outwards. Body somewhat elongate, anterior profile slightly concave above the snout and in front of the orbits, snout protruding, protractile, the lower jaw much the longer; maxilla nearly vertical, equal in length to the diameter of the eye; margin of preopercle inclined forwards, below feebly serrated; spine gently curved, not channeled, reaching to beneath the bony margin of the opercle, a membrane closely invests its whole inner surface; the spine is received into a shallow groove in front of the pectoral. The angle of the opercle is slightly produced, but there is no spine. There is a short recumbent spine immediately in advance of the first erect dorsal spine which arises in advance of the vertical from the opercular border; the eleventh spine is the longest 1·55 in the length of the head and is slightly shorter than the longest rays. The first anal spine arises beneath the eleventh dorsal, to which the third anal is equal, the soft dorsal and anal are not produced but are rounded posteriorly, both terminate in the same vertical line, but the anal rays extend somewhat further back, nearly to the base of the caudal. Pectoral longer than the ventral which latter is contained

1·3 times in the length of the head and the scarcely produced outer rays reach slightly beyond the vent. Caudal rounded, the upper lobe rather the longer, the least height of the pedicel half the length of the head.

Scales.—Head, body and all the fins densely covered with scales, those of the body as seen on the fish are deep and short, strongly ctenoid; they are so small and irregular as to render counting not possible. When removed from the body a single scale is seen to be deep seated with a comb-like margin exhibiting about fourteen teeth ; in *situ* the scales are greatly imbricate, their pectinations alone exposed. The lateral line is to be very faintly traced, it follows the curvature of the back to between the termination of the dorsal and anal fins whence it proceeds along the middle of the caudal pedicel.

Colours.—The head is dirty yellow, (brighter in the younger example) darker on the preopercle. On the occiput the colour gradually merges into that of the body, which together with the bases of the vertical fins and the caudal pedicle is a rich coffee-brown ; the exposed portions of the jaws, the chin and chest are a darker brown. The eye is encircled by a dark line, which in life may have shown traces of blue ; above, behind and below the eye this line follows the curvature of the orbit, but anteriorly is produced forwards, the upper part passing through the nostrils and finally joining the lower portion on the anterior edge of the pre-orbital. Another line marks the posterior edge of the preopercle, is deflected up the base of the spine and returns along its lower border to the point of origin ; the opercle is also margined by a dark line and the opercular membranes are black. Anteriorly a broad submarginal band of brown arises on the membrane of the dorsal spines and passes along the whole of the fin, becoming darker posteriorly; the margin of the membrane is yellow. The whole of the anal is very dark brown and is margined with greyish-blue. The pectorals are yellow basally, the posterior half black with a yellow margin; the ventrals are without markings. The caudal is yellow, which colour is sharply separated from the brown of the pedicel, its upper and lower margins are narrowly edged with black, and its posterior third is also black, this colour being narrow on the lobes and broadening greatly towards the middle ; the rays are narrowly margined with yellow. Total length of two specimens 250 mm. and 170 mm. respectively.

HOLACANTHUS SEMICINCTUS, *sp. nov.*

(Plate xxxvi.)

D. xv. 17.   A. iii. 18.   V. i. 5.   P. 17.   C. 17 + 4.   L. lat. 46.
    L. tr. 7 + 22.

Length of head 3·85, of caudal fin 4·53, height of body 2·1 in the length (caudal excluded). Diameter of eye 3·6, length of snout

3·1, of preopercular spine 2·0 in the length of the head; the inter-orbital space is convex, one-fourth more than the diameter of the eye. Gill-rakers narrow, of moderate length, eleven on the lower limb of the first arch. The teeth are cardiform, arranged in a narrow band in each jaw, each with a tricuspid apex. The body is rather elongate, the anterior profile slightly rounded and tumid above the snout; jaws equal, only slightly protractile, the maxilla is shorter than the diameter of the eye; preorbital produced into a spine directed horizontally forwards and its lower angle into three smaller spines directed forwards and downwards, followed by four small points. Preopercle inclined forwards, its hinder limb strongly serrated, its lower with four strong denticles. The spine is gently curved and received into a shallow groove in front of the pectoral, its point reaches the vertical from the margin of the opercle, a membrane invests its inner surface. The opercle and subopercle finely denticulated the points rather distant. The first dorsal spine is situated above the opercular margin, and its length equals the diameter of the eye, the others gently increase in length to the fifteenth which is the longest, and last, exactly twice the length of the first, the central rays are produced but not filamen-tous, the tenth which is the longest, is three-fourths the length of the head.

The first anal spine arises beneath the eleventh dorsal, the third is the longest, 1·6 in the length of the head; all are stouter than the dorsals. The rays are similar in character and extent to those of the dorsal and it is also the tenth which forms the summit of the fin which terminates evenly with the dorsal. The pectoral is equal to the head in length, as is also the ventral but its first ray is produced, the filament reaching to the first anal spine. Caudal lunate its upper and lower rays produced into filaments, the height of the pedicel is half the length of the head.

Scales.—The scales of the body are large, the exposed portions angular in shape, strongly ciliated; removed from the body each scale is subcircular in outline a little higher than long; the scales on the fins are very small. The lateral line rises to below the fifth dorsal spine whence it follows the curvature of the back to near the termination of the rays, it then bends abruptly down-wards and runs horizontally along the caudal pedicel.

Colours.—After long immersion in spirits the general colour is yellowish, tending to brownish on the head and dorsal surface, the fins are also yellowish and immaculate, with the exceptions below mentioned. The upper half of the body is crossed with eight or nine sub-vertical lines of dark brown, they have a slight posterior inclination and anteriorly do not cross the mid line of the body, they descend lower behind but do not even there reach the ventral profile; the first distinct line arises from the base of the third dorsal spine but there is a faint trace of one in advance of it at the base of the first spine. On the throat is a series of large, pale brown blotches arranged alternately, the dorsal and anal fins are margined by an extremely narrow edge of dark brown and there are two small spots of this colour on the 14 – 15 dorsal rays, one near the base, the other nearer the margin. On the caudal pedicel the dark bands of the body are represented by spots and on the middle caudal rays are five series forming interrupted vertical bands.

Total length of specimen 195 mm.

The species to which our specimen is nearest allied is *H. melanospilus*, Bleeker,[*] (from the sea of Amboyna) which indeed in many particulars it closely resembles. I have not access to the original description, and my conclusions are formed from Günther's epitomised description,[†] and a comparison with Bleeker's figure.[‡]

In *H. melanospilus* the body bands are very close together, eighteen or nineteen in number, and are continued to the ventral surface ; there is a large, round, black, white-edged spot on the thorax, and the vertical fins have small, white rings. In *H. semicinctus* the body bands are widely spaced, do not exceed nine in number, and are not continued below ; there is no prominent spot on the thorax, and the caudal is marked with dark blotches. The horizontal preorbital spine which forms a noticeable feature in our specimen is not so represented in Bleeker's illustration.

ACANTHOCAULUS, gen. nom. nov.

This name is suggested to replace *Prionurus*, Lacépède, 1830, preoccupied by Ehrenberg, in Arachnida, 1829. In 1898, Jordan and Evermann instituted a genus *Xesurus*[||] to receive certain American species, some of which had been described under the name *Prionurus ;* the authors write " This genus is close to the East Indian genus *Prionurus*, Lacépède, differing chiefly in the character of the caudal armature, the plates in *Prionurus* being small, sharper, and in greater number." Messrs. Gilbert and Starks also admit the validity of the genus by adopting it for their species *Xesurus clarionis*. I have therefore no choice, in pointing out that the term *Prionurus* is inadmissible in Ichthyology, but to propose a new name. *Acanthocaulus* will include *P. microlepidotus*, Lacépède, an Australian species, and *P. scalprum*, Langsdorf, from Japan, but *P. maculatus*, Ogilby,

---

[*] Bleeker—Act. Soc. Sci. Indo-Nederl., ii. Amboina, p. 56 *(fide* Günther).
[†] Günther—Brit. Mus. Cat. Fish., ii., 1860, p. 48.
[‡] Bleeker—Atlas Ichth., ix. pl. ccclxviii. fig. 2.
[||] Jordan and Evermann—Rep. U. S. Com. Fish and Fisheries, 1895, p. 421, (name only); and Bull. U. S. Nat. Mus., No. 47, 1898, p. 1694 (description).

from Port Jackson and Lord Howe Island will have to be referred to *Xesurus.*

XESURUS MACULATUS, *Ogilby.*

Prionurus maculatus, Ogilby, Proc. Zool. Soc. 1887, p. 395.

At separate times three examples of this species were received from the island; they were included in gatherings by Messrs. Langley and Icely. Although the type was described from a specimen taken on the coast of New South Wales, it is more than probable that the species is commoner around the island than on the shores of the mainland.

NASEUS UNICORNIS, *Forskal.*

Two examples have been received, one in March, 1891, from Mr. Langley, and the other in August, 1899, from Mrs. T. Nicholls. Macleay has recorded the species from Torres Straits.

ALUTERA MONOCEROS, *Osbeck.*

The known distribution of this species is now considerably extended, Mr. Icely having in 1894 obtained two examples from the Island. It had not been previously recognised from Australia. It has also been recently added to the known fauna of the United States by Dr. Hugh M. Smith, in reference to which Dr. D. S. Jordan publishes an interesting notice.* He therein expresses the opinion that the example represents a species distinct from *A. monoceros.*

OSTRACION CUBICUS, *Linnæus.*

We have received, at various times, a number of specimens from the Island. Although not recorded from Eastern Australia, this species has been taken on its northern and southern coasts.

AMBLYRHYNCHOTUS OBLONGUS, *Bloch.*

The only adult example I have seen from the Island was brought by Mr. Icely in 1895, but there are two small examples of older date in the collection which are possibly also referable to *A. oblongus.* The species was not recorded from the east coast of the mainland until 1898, when it was obtained by the Thetis Expedition.

OVOIDES MELEAGRIS, *Lacépède.*

Tetrodon meleagris, Lacépède, Hist. Nat. Poiss., 1798, i., p. 505.

In January, 1895, Mr. Icely brought us a small "Tetrodon," which I have identified with this Polynesian species.

---

* D.S.J.—American Naturalist, xxxiv., 1900, p. 69.

EUCHILOMYCTERUS, *gen. nov.*

The single specimen, regarded as the type of a new genus, differs from the descriptions of all described species of *Chilomycterus*, by having the anterior dorsal and a temporal spine four-rooted. Owing to the indifferent condition of the specimen, the nature of the nasal tentacles (if present) cannot now be determined; the presence of a spine in the middle of the forehead would indicate a nearer subgeneric affinity to *Cyclichthys* than to *Chilomycterus*. As far as ascertainable, therefore, the new genus may be characterised as follows:—Body broad, compressed?. Dermal spines very short, immovable, mostly with three roots, the anterior dorsal, and a temporal spine, with four roots ; all roots (excepting those of the temporal one) overlap, forming a coat of mail. Caudal practically without pedicel; fins small; jaws without median suture.

EUCHILOMYCTERUS QUADRADICATUS, *sp. nov.*

D. 12.   A. 10.   P. 17.   C. 9.

Length of head 2·8, of caudal fin 3·7 in the total length, interocular space flat and broad, the distance between the middle supraocular spines equal to the length of the head, eyes large, lateral, not quite so long as the snout. The dorsal spines are very low, little more than tubercles: eight or nine between the eye and the tail, those on the sides and belly are scarcely larger, excepting the posterior ones which do not however attain to 3 mm. in length ; the two lateral spines behind the dorsal fin throw their inner roots together on the median line, and behind them is a pair of roots bearing a minute median spine. With this exception, and those below noticed, all the spines are three rooted, the anterior root covering the lower lateral one of the spine in advance of it. All the roots form strong butresses against the spine. Three supraorbital spines, the anterior of which sends its inner root across the forehead, and projected between them is the anterior root of a minute median spine. From this point, backwards, to between the gill openings, the spines have four roots arranged longitudinally and transversely. A four-rooted spine (the temporal) exists between the the third supraocular and the large three-rooted spine situated above the gill opening ; it is of regular shape, and its roots scarcely touch those of any other spine. The coloration cannot be given, as the specimen is dry and faded. Total length 180 mm.

I have not figured this species because its most striking features have already been sufficiently expressed and the specimen is not in suitable condition for delineation. It was forwarded to us by Mrs. Nichols in January last. Apart from the peculiarities mentioned, it seems to have some resemblance to *Chilomycterus tigrinus*, Cuvier.

PTEROIS ZEBRA, *Cuvier and Valenciennes.*

P. volitans proves to be quite common off the Island whence we have received many examples. The inclusion of *P. zebra*, on the other hand, rests on the evidence of a single specimen forwarded by Mr. Leely. Both species occur in Port Jackson, but, similarly, the former is much the commoner, or, as I should say, the less rare. Neither species has been recorded on the Australian Coast south of Port Jackson.

PARAPERCIS CYLINDRICA, *Bloch.*

Sciæna cylindrica, Bloch., Ichth., 1797, pl. 299, fig. 1.

We possess two examples from the Island, one obtained by Mr. Leely in 1893, and the other by Mrs. Nicholls in January last. Günther[*] mentions that a specimen from Aneitoum differs from others in the British Museum by having fifty-eight, instead of fifty transverse rows of scales, and by the ventral fin reaching only to the third, instead of the fifth anal ray. In both our specimens there are fifty-four scales in the lateral line; in one, the ventrals reach to the third anal ray, in the other, to the first only. In an example from the Admiralty Islands there are also fifty-four series of scales, while the ventrals reach to the fourth anal ray.

CRISTICEPS AUSTRALIS, *Cuvier and Valenciennes.*

Of the two species originally recorded from the Island, namely, *C. aurantiacus*, Cast., and *C. rosens*, Günth, the latter enters Ogilby's genus, *Petraïtes*. A specimen secured by Mr. Leely is to be identified with *C. australis*, Cuv. and Val. Port Jackson examples of which Castelnau designated *C. macleayi*.

---

[*] Günther—Brit. Mus. Cat. Fish, ii., 1860, pp. 239 and 525.

NOTES on FISHES from WESTERN AUSTRALIA, and DESCRIPTION of a NEW SPECIES.

By EDGAR R. WAITE, F.L.S., Zoologist.

(Plate xxxvii.)

By an arrangement with Mr. B. B. Woodward, Curator of the Perth Museum, we have received a small collection of fishes obtained, for the most part, in the Swan River, near Perth.

No attempt has been made to catalogue the fishes of the western coast of Australia, and indeed, with the exception of isolated records, little has been done since the early voyagers collected there. We have Castelnau's "contribution,"* and the following species received by us are recorded by this writer from the neighbourhood of Freemantle.

DOROSOMA EREBI, *Günther*.

HEMIRHAMPHUS INTERMEDIUS, *Cantor*.

As *H. melanochir*, Cuvier and Valenciennes.

SPHYRÆNA NOVÆ-HOLLANDIÆ, *Günther*.

In addition to this species and *S. obtusata*, it is possible that we may have a third species in New South Wales, for the description of *S. novæ-hollandiæ*, by Ogilby,† does not tally with that form. In typical examples the ventral is inserted wholly in advance of the first dorsal, while that writer describes the fin as being inserted beneath the anterior half of the first dorsal ; he, however, figures it (Pl. xxx.) more in agreement with our examples.

THERAPON ELLIPTICUS, *Richardson*.

Mr. Woodward informs us that the specimens forwarded were taken at Kimberley, in fresh water.

THERAPON CAUDAVITTATUS, *Richardson*.

SPAROSOMUS AURATUS, *Bloch and Schneider*.

Noticed by Castelnau under the synonym *Pagrus unicolor*.

PLATYCEPHALUS LÆVIGATUS, *Cuvier and Valenciennes*.

In addition to the foregoing and the introduced :—

CARASSIUS AURATUS, *Linnæus*, and

CARASSIUS CARASSIUS, *Linnæus*,

---

* Castelnau—Proc. Zool. Soc. Vict., 1873, ii., pp. 123 - 149.
† Ogilby—Edible Fishes N.S.W., 1893, p. 114.

we have received the following species :—

GONORHYNCHUS GREYI, *Richardson.*

TYLOSURUS FEROX, *Günther.*

I am not aware that this species has been previously recorded from West Australia. Castelnau has described a species under the name *Belone gavialoides,** which, judging from the description, and taking into account the relative position of the fins, is distinct from *T. ferox.*

TRACHURUS DECLIVIS, *Jenyns.*

COLPOGNATHUS DENTEX, *Cuvier and Valenciennes.*

Günther's type of *C. richardsonii,*† regarded by Boulenger‡ as synonymous with *C. dentex,* was obtained at Freemantle, at the mouth of the Swan River, wherein Mr. Woodward's specimens were taken. Our examples are without markings of any description, and this fact, taken in conjunction with the widely different colouration or ornamentation of the figures of Quoy and Gaimard,§ Richardson,‖ and Günther,¶ indicates that the species is subject to great variation in colour and pattern.

CHRYSOPHRYS DATNIA, *Forsk.*

Although this species does not appear to have been previously noticed from Western Australia, it was naturally expected that, having such an extensive distribution, it would sooner or later be thence recorded. It may now be said to occur on the whole of the eastern, northern, and western seaboards, but being so much more numerous in the tropics, we are scarcely likely to find more than a straggler or so on our southern shores.

ODAX RICHARDSONII, *Günther.*

The specimens received do not differ from examples taken in Port Jackson; the dark markings on the body are very pronounced, and in this the examples are not unlike *O. semifasciatus,* Cuv. and Val., from which the species is distinguished by the serrated preoperculum and by the smaller number of scales above the lateral line—seven in *O. richardsonii,* fifteen in *O. semifasciatus.*

The serrations in some specimens are so slight as to be of doubtful specific value, yet Castelnau proposed for examples with serrated preoperculum, the generic name *Neodax.*

---

* Castelnau—Proc. Zool. Soc. Vict., ii., 1873, p. 142.
† Günther—Proc. Zool. Soc., 1861, p. 391.
‡ Boulenger—Brit. Mus. Cat. Fish, (2), i., p. 310.
§ Quoy and Gaimard—Voy. "Astrolabe," Poiss., pl. iv., fig. 2.
‖ Richardson—Zool. Ereb. and Terr., Ichth., pl. lvii., figs. 3 - 5.
¶ Günther—*Loc. cit.,* pl. xxxviii.

PERIOPTHALMUS KOELREUTERI, *Pallas.*

The occurrence of this species in such a comparatively high latitude as Perth, is another instance of the more tropical character of the west than the east coast of Australia. On the east, Castelnau records it* from the entrance of the Brisbane River, south of which it has not been observed. *P. australis,* Cast., is said to be found on the mud flats of the Richmond River, New South Wales.† Saville Kent, in his paper on the Marine Fauna of Houtman's Abrolhos Islands,‡ shows how the Abrolhos support a wealth of tropical life, such as Holothurians and the more brilliantly coloured Labroids, familiar to him from Torres Straits and the more northern regions of the Great Barrier Reef.

Houtman's Abrolhos are of coral growth, a formation met with on the eastern mainland only in much lower latitudes, and in explanation Mr. Saville Kent writes:—" The anomalous character of the marine fauna of Houtman's Abrolhos, as herein defined, can only be accounted for by the assumption that an ocean current, setting in from the equatorial area of the Indian Ocean, penetrates as far south as this island group, and has borne with it the floating embryos of the Holothuridæ and Cœlenterates, etc., that so characteristically distinguish it. A reference to the Admiralty charts, dealing with the ocean currents of this region, supports this interpretation to a considerable extent; indicating as a matter of fact, a prevailing northerly set along the western coast of Australia, but at the same time a distinct southerly intrusion of the waters of the Indian Ocean at some distance off shore, down towards and closely approaching Houtman's Abrolhos."

HOPLEGNATHUS WOODWARDI, *sp, nov.*

(Plate xxxvii.)

B. v.   D. xi. 11.   A. iii. 2.   V. i. 5.   P 17.   C. 17.   L. l. 62.
        L. tr. 25 - 60.

Length of head 2·57, of caudal fin 5·76, height of body 2·31, in the total length (caudal excluded). Eye very large, 3·63 in the length of the head, 1·25 in the snout, and 1·13 in the interorbital space, which is slightly convex. Nostrils approximate, the anterior round, the posterior elongate, its own length in advance of the margin of the eye. The upper profile of the head with a pronounced swelling above the anterior nostril, and forming a sharp bony keel on the occiput. Dorsal and ventral profile a gentle curve. Upper jaw the longer. Cleft of mouth medium, almost horizontal, the maxilla extending to within the anterior margin

---

* Castelnau—Proc. Linn. Soc. N.S.W., ii., 1878, p. 231.
† Ten.-Woods—Fish and Fisheries N.S.W., 1882, p. 27.
‡ Saville Kent—Rep. Brit. Assoc., 1895, p. 732.

of the orbit. Opercles entire, with one flat jagged spine. Post-temporal and clavicular plates very pronounced.

Teeth.—These consist of a bony lamella in each jaw, with median division, as in *Tetrodon*; the lamella is translucent, and the summit of each tooth can be traced in its substance, the whole forming a regular diagonal mosaic. As the teeth are successively pushed to the margin of the lamella, their crowns become free and they then form a sub-imbricate series, each crown being grey, tipped with black. These peculiarities are more noticeable in the lower than in the upper jaw. Behind the anterior series is a group of rounded teeth, white in colour; within the upper lateral series are a few isolated teeth, similar in colour and form to those in the lamellæ.

Dorsal spines very strong, compressed, increasing in height to the seventh, which is exactly half the length of the head, and higher than the rays ; the last spine nearly equals the fourth in length; the basal length of the spinous is nearly twice that of the soft portion. The anal spines are rather stronger than those of the dorsal, the third somewhat exceeds the second in length, and is 2·6 in the length of the head, and equals the fourth dorsal ; the rays are similar to those of the rayed dorsal. The fourth upper ray of the pectoral is the longest, it is rather longer than the ventral, and is contained 1·7 times in the length of the head. The ventral spine is similar to the longest dorsal in character and extent, and the fin all but reaches the vent. The caudal is emarginate, the upper lobe slightly the longer ; the least height of the pedicle is one-third the length of the head. The spinous portions of the dorsal and anal fins are received into a deep groove, and the soft portions are scaly at the bases, as is also the caudal.

Scales—small, finely ctenoid or ciliate, those on the opercules freer and of more angular contour than those of the body. Upper part of head, snout, maxilla, mandible, and two or three elongate areas above and behind the eye naked, otherwise scaly.

Colours—Yellowish or brownish, which may in life have been pink. The fins are dusky and without markings, excepting the dorsal and anal, which are blotched, as below described. The markings on the body are five broad black vertical bars. The first passes from the top of the head, through the eye, and down the cheek. The second arises in advance of the dorsal fin, involving the first two spines, thence across the base of the pectoral. The next bar passes from the 7 – 10 dorsal spines to the vent. The fourth connects the dorsal and anal rays, forming a black blotch on each fin, and continued backwards along the base of the anal rays ; while the fifth, which is narrower, passes across the base of the caudal pedicel. All the bars are inclined obliquely backwards and are narrower towards the ventral surface.

This species is perhaps the one doubtfully referred by Johnston[*] to *H. conwayi*, Rich.[†] It is, however, quite distinct from that species, and differs in the following particulars :—

The dorsal has a smaller number of spines, and is relatively very much higher, the spines also are longer than the rays; whereas in *H. conwayi* the rays are twice as long as the spines— a character common also to the anal.

H. fasciatus and *H. punctatus*[‡] have each twelve dorsal spines and sixteen rays, the body in these species is shorter and higher, the eye is smaller and the soft vertical fins much longer than in *H. woodwardi*.

In common with other Australian workers, I have at times referred to the difficulty experienced by zoological writers living at prohibitive distances from European literary centres, and the hopelessness, in many cases, of bringing an undertaking to a satisfactory conclusion. Such disabilities are caused by a lack of necessary literature, and many are the instances in which a train of research has to be abandoned owing to the impossibility of consulting some particular paper.

Where a genus is weighted with a large number of species, the difficulty may be appreciated ; but when only a few are known, the task would seem to be a simple one ; this may not, however, be so, and I may instance *Hoplegnathus*, the genus now under consideration.

Richardson first described the genus in 1840, as a Scaroid, under the name *Oplegnathus*,[§] the species being *O. conwaii*. The following year he altered the generic name to *Hoplegnathus* and the specific one to *conwayi*, when exhibiting drawings before the meeting of the British Association,[‖] and in 1849 published a full description and figure.[¶] The specimen described was supposed to be from Australia. In the year 1844, Temminck and Schlegel described two fishes from Japan under the generic name *Scarodon*, namely, *S. fasciatus* and *S. punctatus*,[‡] and mention the earliest representation of a species in the Atlas of Krusenstern's voyage, under the name "Poisson perroquet noir."[**]. Both these examples were from Japan.

---

* Johnston—Proc. Roy. Soc. Tas., 1881, p. 191.
† Richardson—Proc. Zool. Soc., 1840, p. 27; and Trans. Zool. Soc., 1849, iii., p. 144, pl. vii., fig. 1.
‡ Temminck and Schlegel—Fauna Japon, Pisces, 1844, p. 89, pl. xlvi. and p. 91.
§ Richardson—Proc. Zool. Soc., 1840, p. 27.
‖ Richardson—Rep. Brit. Assoc., 1841 (1842), pt. 2, p. 71.
¶ Richardson—Trans. Zool. Soc., iii., 1849, p. 144, pl. vii., fig. 1.
** Krusenstern—Atlas, pl. lii., fig. 2.

We next turn up Richardson's paper on the Ichthyology of the seas of China and Japan,* and find that he recognised the generic identity of *Scarodon* with his own *Hoplegnathus*, and under *H. punctatus* mentions having seen, very cursorily, in the Museum at Fort Pitt, a spotted *Hoplegnathus* from Norfolk Island. As the only island of that name, according to the atlas and the gazetteer, is the dependency of New South Wales, it would seem as though this species should be credited to our fauna, but Richardson describes its habitat merely as the seas of Japan and China. In the work quoted, he, with doubtful judgment, coins a third name—*H. maculosus*, his type being a drawing only, at the same time he doubts its specific distinction from *H. punctatus*.

In 1854, Bleeker raised the genus to family rank under the name *Hoplegnathoidei*,† but I have not access to his paper ; he again mentions it in his Archipelago Indico.‡ Two years later, Richardson, who had apparently not seen Bleeker's work, placed his *Hoplegnathus* as a genus, under *Chætodontidæ*.§ The three valid species mentioned, are recorded by Günther‖ in 1861, but it becomes evident that one paper on the subject had at that time been overlooked, of which more later.

On referring to the Zoological Record for 1865, we read¶:— "*Hoplognathus*. M. Guichenot states that *Ichthyorhamphus* (Casteln.) from the Cape of Good Hope is identical with this genus. Mém. Soc. Sc. Nat. Cherbourg, xi., p. 5. The same author refers it to the Scaroid fishes ; but its pharangeal bones are entirely separate, rather feeble, and armed with villiform teeth." The work in which Guichenot published the observation is not accessible to me, and I am unable to find where Castelnau's genus was described. It is omitted from the "Nomenclator Zoologicus" of Scudder, and on searching the Royal Society's Catalogue such references as I can consult do not contain notice of the genus *Ichthyorhamphus*, so that I am unable to learn even the specific name applied by Castelnau.

The following reference is supplied by the Zoological Record for 1867 :—**"*Hoplognathus fasciatus* (Kröy.) is described as *Scarostoma insigne* (g. et sp. n.) by Prof. Kner, Sitzgsber. Ak. Wiss. Wien, 1867, lvi., p. 715, fig. 3," and the same subject is recorded in the Zoological Record for 1868, as follows††:—"Prof. Kner also has recognised the identity of his *Scarostoma* with this genus (See

---

* Richardson—Rep, Brit. Assoc., 1815, p. 247.
† Bleeker—Ver. Akad. Wetensch. Amsterdam, i., 1854, Japan, p. 6.
‡ Bleeker—Spec. Pisc. Arch. Indico, 1859, p. 250.
§ Richardson—Encyc. Brit. (Ed. ix.), Ichth. xii., p. 303.
‖ Günther—Brit. Mus. Cat. Fish., iii., 1861, pp. 357 - 8.
¶ Günther—Zool. Record, 1865, Pisces, p. 184.
** Günther—Zool. Record, 1867, Pisces, p. 161.
†† Günther—Zool. Record, 1868, Pisces, p. 146.

Zool. Record, iv., p. 161); but he still thinks that the fish described by him is a new species (Wiegm. Arch. 1868 in Troschel's Bericht). [It is *Hoplognathus fasciatus* of Kröyer, not of Schlegel; the name of the Japanese species may be changed to *Hoplognathus krusensternii*.]" We do not possess the Vienna publication, so that further research in this direction is impossible. There is no reference to where Kröyer's paper was published, but such is ultimately traced by Carus and Engelmann's Bibliotheca Zoologica (1861, p. 1028); the reference being:—"*Oplegnathus fasciatus.* in: Kröyer, naturhist. Tidsskr. N. R. Bd. i., 1845, p. 213 – 223," a work to which again I cannot refer. In passing it may be noted that the Bibliotheca does not record Castelnau's *Ichthyorhamphus*.

I have no direct evidence as to where Kröyer's type was obtained, but Günther writes of the family *Hoplognathidæ*[*]:— "One genus only is known, *Hoplognathus*, with four species from Australian, Japanese, and Peruvian coasts": as we know the species representing the two former habitats, I presume Kröyer's example was from Peru, and it is possible that *H. woodwardi* is identical with *H. fasciatus* from Peru, many types being common to Australia and South America. It is to be noticed that the Cape of Good Hope, supposed to be represented by *Ichthyorhamphus*, is not included in the distribution of the family.

Although the Fauna Japonica, Pisces, bears on the title page the date 1850, the work was issued in parts, commencing 1844, in which year the decade containing *Hoplegnathus* appeared. It thus antedated Kröyer's paper, published in 1845, which was however not recorded by Günther in his Catalogue, and this constitutes the omission previously referred to. The changing of the name of the Japanese species was therefore not justified, as acknowledged later by using *H. fasciatus*, according to priority.[†] Steindachner has redescribed the species, but unfortunately I am unable to consult his paper.[‡]

In changing the spelling of *Hoplɛgnathus* to *Hoplognathus*, Günther had apparently assumed that the derivation of the prefix was ὅπλον = ARMA, whereas Richardson expressly states that his derivation was ὁπλή = UNGULA.[§]

Further, the name *Hoplognathus* is inadmissible for this genus, having been used in 1819 by MacLeay, and again by Chadoir in 1835, for different genera of Coleoptera. It was subsequently (1844) used by Burmeister, also in Coleoptera.

---

* Günther—Study of Fishes, 1880, p. 410.
† Günther—Challenger Reports, Zool., i., Shore Fishes, 1880, p. 64.
‡ Steindachner—Sitz. K. Akad. Wiss. Wien., cii., 1893, p. 222.
§ Richardson—Trans. Zool. Soc., 1849, iii., p. 144.

The CARD-CATALOGUE SYSTEM ADAPTED TO Museum Requirements.

By Edgar R. Waite, F.L.S., Zoologist.

In a thoroughly up-to-date Museum there must always be going on an active exchange of specimens with kindred institutions in other countries. To catalogue the collections in such an establishment may in itself be a matter of some difficulty. If one is content merely to enter the names and particulars of current acquisitions in a book form register, and rule out, or otherwise mark, entries representing specimens sent away, nothing could be simpler. Such a register, however, cannot be kept in systematic order; a great disadvantage when dealing with Natural History specimens, and hours may be spent in tracking the source of any particular object.

In dealing with the large number of specimens under my care at the Museum, namely, Mammals, Reptiles, Fishes, and all Osteological preparations, I had the inadequacy of the usual form of register for ordinary working purposes, forcibly brought home to me; for my own convenience, therefore, I duplicated the record of current donations, etc., according to the plan below referred to.

Eighteen months ago the Curator instructed me to prepare a catalogue of the duplicate Mammals available for exchange, and for this purpose I was provided with an additional register. I then explained what system I had instituted, and the Curator heartily approving, permission was accorded me to officially adopt it in the Institution, as referred to in his Annual Report for 1898.*

The Curator's remarks were based on a six months' trial, during which time a comparatively small catalogue only had been prepared. All the collections in the various sections previously mentioned are being catalogued on this plan, and so far the work has occupied an assistant nearly the whole of the eighteen months indicated.

Many important libraries are now catalogued by the "card" system, and it is simply an adaptation of this to Museum requirements that I desire to bring into notice. Once a book is placed in a library it usually remains there, and if worn out is merely replaced, the substituted book bearing the reference number of the discarded one; changes occur only by interpolating new volumes. With a museum collection the case is different, for, in addition to the new material, specimens are constantly being removed by exchange, and old examples can never be actually replaced, for unlike a book, each has an individuality of its own, depending on locality, age, sex, season, or other condition.

* Aust. Mus. Ann. Report, 1898 (1899), p. 6.

A museum catalogue should be adapted to include the following:

I. Exhibited Collection.

a. Valuable specimens, or single representatives of a species, not necessarily in good condition.

b. Permanent specimens which may reasonably be supposed to be the best procurable.

c. Indifferent specimens retained only until better examples are procured.

II. Duplicate Collection.

d. Reserve—Specimens not at present required for exhibition, but too rare to be parted with.

e. Store—Specimens available for exchange.

III. Type Collection, if not exhibited. It is also necessary to indicate whether the specimens are mounted, in skins, or are preserved in fluid.

As implied by the name, the system consists of indexing by means of loose cards instead of by the ordinary book method. A card is issued for every individual specimen, and upon it written the name of the object and all information concerning it; it is in fact a copy of the collector's ticket, together with the registration and other marks, as Gallery, Duplicate, Type. These cards stand on edge in drawers specially constructed to receive them, and may be arranged in any way desired: the height of the card is less than that of the drawer, so that a deeper series may be inserted, these latter, standing up above the others, are to receive the names of the Orders or Families, etc., and may be of distinctive colour. When properly placed, a card or series of cards may be inserted anywhere or a similar series withdrawn without disturbing the general arrangement.

This system, as I have applied it, is not intended to take the place of the ordinary register, but rather to be a key to the collections : the register would record all specimens received in chronological order, but the changes made in the collections would be indicated by the card catalogue.

I have here done no more than indicate the system, for the arrangement of a catalogue depends so much on the nature and number of the specimens dealt with, and the fancy of their custodian. Personally, I have adopted two cabinets for each section, one to contain the cards of the exhibited, and the other those of the duplicate collections.

For full information as to the working of the card system as used for library purposes, the publications of the Boston Library Bureau (U.S.A.) and of the Manchester Museum may be consulted.

OCCASIONAL NOTES.

V.—*TURRICULA SCALARIFORMIS*, TEN.-WOODS—ITS OCCURRENCE in NEW SOUTH WALES.

On a recent visit to Gerringong, the Curator collected a number of small dead shells from between tide marks on a sandy beach. Among them I have detected a single, rather worn example of *Turricula scalariformis*, Ten. Woods,* a species which has not been previously recorded from this Colony, though observed in South Australia, Tasmania, and Victoria. The synonomy and bibliography of it has been recently reviewed by Messrs. Pritchard and Gatliff.†

The more prominent ribbing and small size, 7 mm., incline the Gerringong specimen to the variety named *Mitra legrandi* by Ten. Woods. For the better recognition of this variety, I add an enlarged drawing of an authentic Tasmanian specimen kindly lent me by the Rev. H. D. Atkinson.

CHARLES HEDLEY.

VI.—*SCALA REVOLUTA*, HEDLEY—ITS OCCURRENCE in FIJI.

In discussing the fauna of Funafuti, I took occasion to point out that the high proportion (one-sixth) of novelties described should be regarded rather as an indication of how little is known of the Pacific fauna than of any native peculiarity. My anticipation that all the minute shells described as new from Funafuti would ultimately be found in the western continental islands is receiving more full and prompt confirmation than I could have hoped for.

Cœcum vertebrale has recently been discovered at Ouvea, Loyalty Islands.‡ And I now have the pleasure of recording *Scala revoluta* from Fiji. A specimen of the latter has been received by the Trustees from Mr. Allan R. McCulloch, who recognised the species in a parcel of sand from Suva.

CHARLES HEDLEY.

VII.—*PHYLLOTHECA* and *CINGULARIA*.

In 1895, I described § a very peculiar plant from the Upper Coal-Measures at Newcastle, that appeared in some measure to unite the characters of the genera *Phyllotheca*, Brong., and *Cingularia*, Weiss. The stems (or branches), stem-discs, leaves and leaf-sheaths, all presented the characteristic features of the former,

* Ten.-Woods—Proc. Roy. Soc. Tas., 1875 (1876), p. 140.
† Pritchard and Gatliff—Proc. Roy. Soc. Vic., xi., 2, 1899, p. 189.
‡ Journ. of Conch., ix , 1899, p. 219.
§ Rec. Geol. Survey N.S.W., iv., pt. 4, 1895, p. 148.

but were accompanied by peltate infundibuliform organs, similar
to those of the latter. These spring from other sheaths on the
stems or branches, like the leaves of an ordinary *Phyllotheca*.
Mr. W. A. Cuneo, of Thirlmere, recently presented to the
Trustees a slab of shale from the Upper Coal-Measures of Shea's
Creek, a branch of the Natti River, in Parish Killiwarra, Co.
Camden, covered with *Glossopteris* leaves, remains of a *Phyllotheca*, as we know it here, and the peltate organs of *Cingularia*,
just as the original specimens were found at Shepherd's Hill,
Newcastle, by Mr. J. B. Henson. This may be considered a very
interesting re-occurrence of a plant that we have yet to learn the
entire structure and full significance of.

The only point of difference that I can detect is a greater length
of the tooth-like projections of the peripheries of the peltate organs
in Mr. Cuneo's specimen, and possibly a less subdivision of these
same parts.

If one of the conclusions I formerly arrived at—that "this plant
seems to be closely allied, if not identical with *Phyllotheca hookeri*,
McCoy," should ultimately prove to have any weight, then possibly
the present examples afford evidence of another species of *Phyllotheca*, possessing peltate infundibuliform organs.

R. ETHERIDGE, JUNR.

VIII.—*LYGOSOMA FRAGILE*, GÜNTHER.

The Trustees have recently received from Mr. Alfred Stanley
Read two small Lizards, which I identify with *Lygosoma (Rhodona)
fragile*, Günther.[*] This species does not appear to have been
recorded since first described from the Peak Downs (Clermont),
Queensland. The new locality is Angledool, in the county of
Narran, New South Wales, just outside the Queensland border,
and four hundred and sixty miles almost due south of the Peak
Downs.

Mr. Read while remarking that the lizard is very rare in the
district, makes some interesting observations on its habits ; he
states that it is never seen on the surface, all found having been
taken from six to nine inches under ground, always in sandy soil.
They are generally turned up at the roots of small stumps, when
they move just like a snake.

The lizards were forwarded alive, we therefore had the opportunity of verifying Mr. Read's observations as to their movements,
and found that when passing through the soil, an action per-
formed with great celerity, the degenerate limbs are closely
adpressed to the body and are not used in subterranean progression.

EDGAR R. WAITE.

---

[*] Günther—Journ. Mus. Godeffroy, xii., 1876, p. 45.   Boulenger—Brit.
Mus. Cat., Lizards (2) iii., 1887, p. 334, pl. xxvii., fig. 2.

EXPLANATION OF PLATE I.

Fig. 1. Stone implement profusely ruddled, and with a large central incised spiral, above it two small discs of concentric circles, the whole enclosed above and below by transverse cross-bar incisions.

,, 2. The opposite face of the same, with a submodian disc of concentric circles surmounted by a three-quarter disc of concentric circles fainter than the former. Faint cross-bars are visible below the submedian disc.

,, 3. "Bull-roarer," Urania Tribe, Linda Creek, W. Queensland.— Five discs, the central ones separated from the others by a cross-bar above and below.

,, 4. The opposite face of the same. There are no cross-bars here, but beneath the terminal discs are three-quarter circles facing towards one another.

Figs. 1 & 2.

Fig. 3.

Fig. 4.

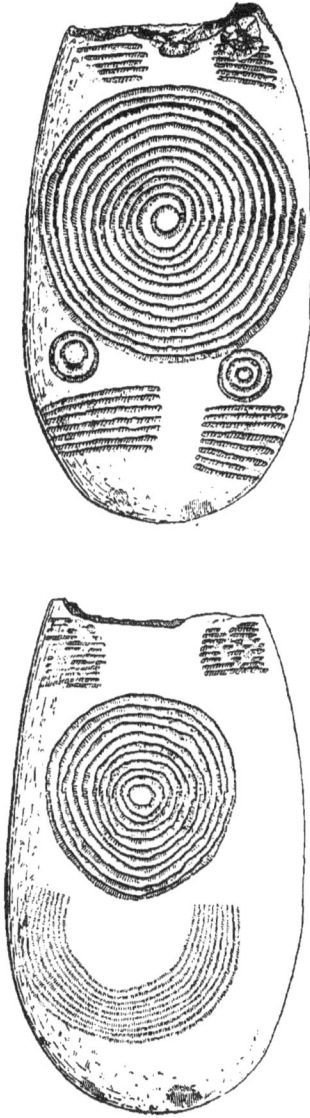

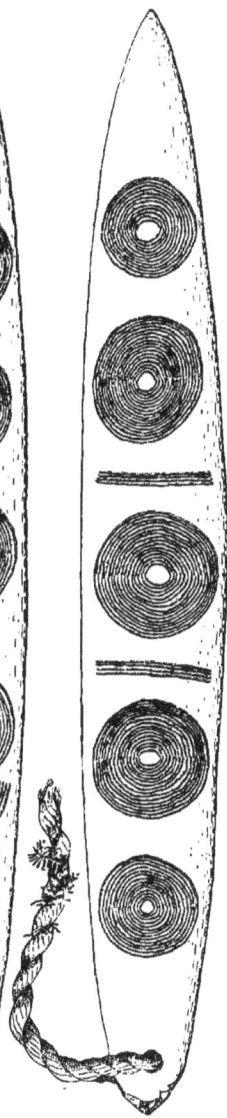

Fig. 5. "Bull-roarer," said to be from S. Australia. A central figure consisting of circumferential circles, surrounding a nucleus of circles, and guarded above and below with the three-quarter circular figure as before.

   ,, 6. Opposite face of the same, with indiscriminately scattered small circles and semi-circles in various degrees of completeness.

   ,, 7. Small "Bull-roarer," also said to be from S. Australia, incised with circles and cross-bars, after the pattern of Fig. 3, except that the outer circles of incisions are incomplete, rendering them in fact semi-circles.

   ,, 8. Opposite face of the same, with a serpentine figure looped on itself, margined by bow-shaped incisions, and the re-entering angles occupied by short transverse bars.

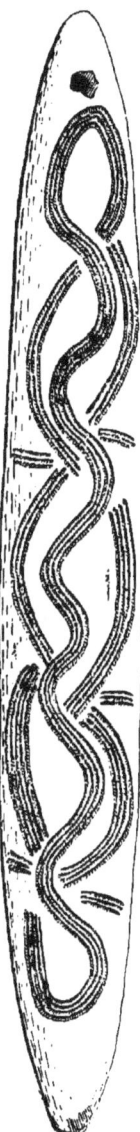

Fig. 5.   Fig. 6.   Fig. 7.   Fig. 8.

EXPLANATION OF PLATE III.

Actinoceras hardmani, Eth. Fil.

Fig. 1. Lateral view showing the septa, and beaded siphuncle at the distal or younger end.

,, 2. Beaded siphuncle— × 2.

,, 3. Section of the siphuncle showing involutions of the membrane— × 3.

,, 4. Pyriform involutions at the proximal or older end of the shell, seen in weathered section— × 5.

,, 5. Involutions and intermediate ridges at the proximal or older end of the shell, seen in partially oblique weathered section— × 5.

Nest and eggs of *Anthus australis*, the Australian Pipit or "Ground Lark."

(The nest is constructed inside a rusty tin.)

EXPLANATION OF PLATE V.

Fig. 1. Cervical vertebra, anterior view.
,, 2. ,, ,, posterior ,,
,, 3. ,, ,, neural ,,
,, 4. ,, ,, hæmal ,,
,, 5. ,, ,, lateral ,,
,, 6. Phalange, side view.
,, 7. ,, more elongated than Fig. 6, side view.

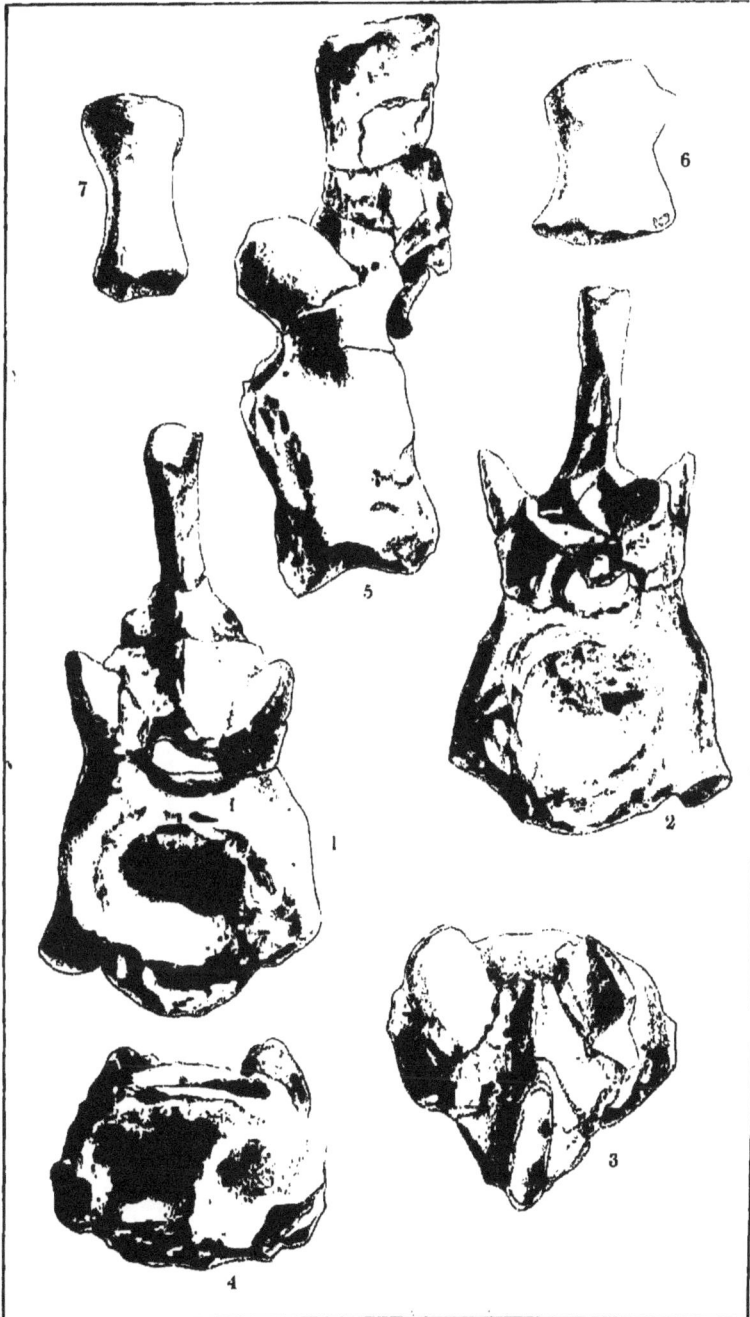

EXPLANATION OF PLATE VI.

Fig. 1. Cervical vertebra, anterior view, with proximal ends of both ribs attached.

,, 2. Two cervical vertebræ in apposition, seen from the right side; the single fractured costal facet on the first is well displayed.

,, 3. The same vertebræ from the left side.

,, 4. Portion of a rib.

,, 5. Distal end of a diapophysis of a dorsal vertebra ?

,, 6. Paddle-bone, possibly the intermedium ? side view.

,, 7. The same, end view.

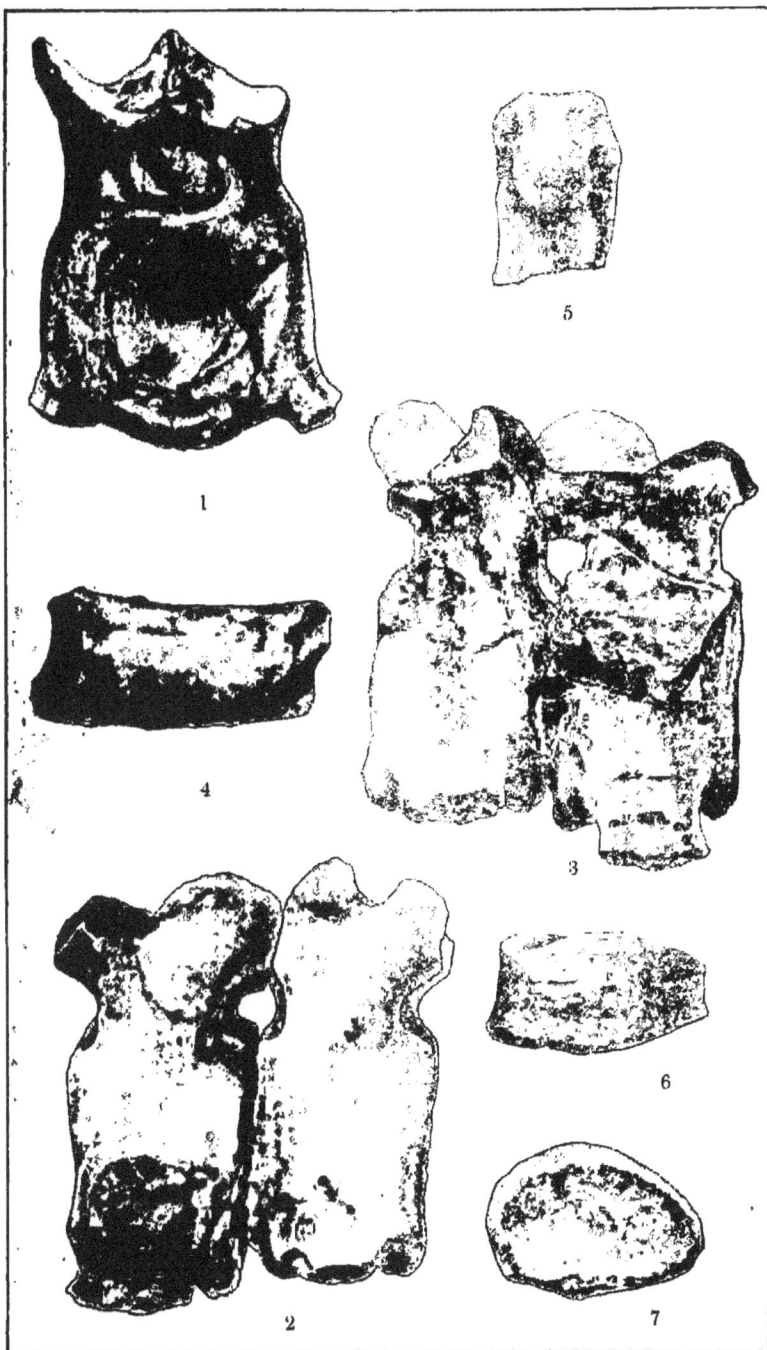

5

1

4

3

2

6

7

Frank R Leggatt Del.

EXPLANATION OF PLATE VII.

Fig. 1. Humerus, proximal end.

„ 2. Distal end of a cervical rib, or "hatchet-bone"?

„ 3. Distal end of a diapophysis of a dorsal vertebra?

„ 4. Portion of the distal end of a trunk rib.

„ 5. Proximal end of a trunk rib.

„ 6. Cup-shaped end of Fig. 5.

„ 7., 8., 9. Phalanges, side views.

„ 10., 11. Teeth.

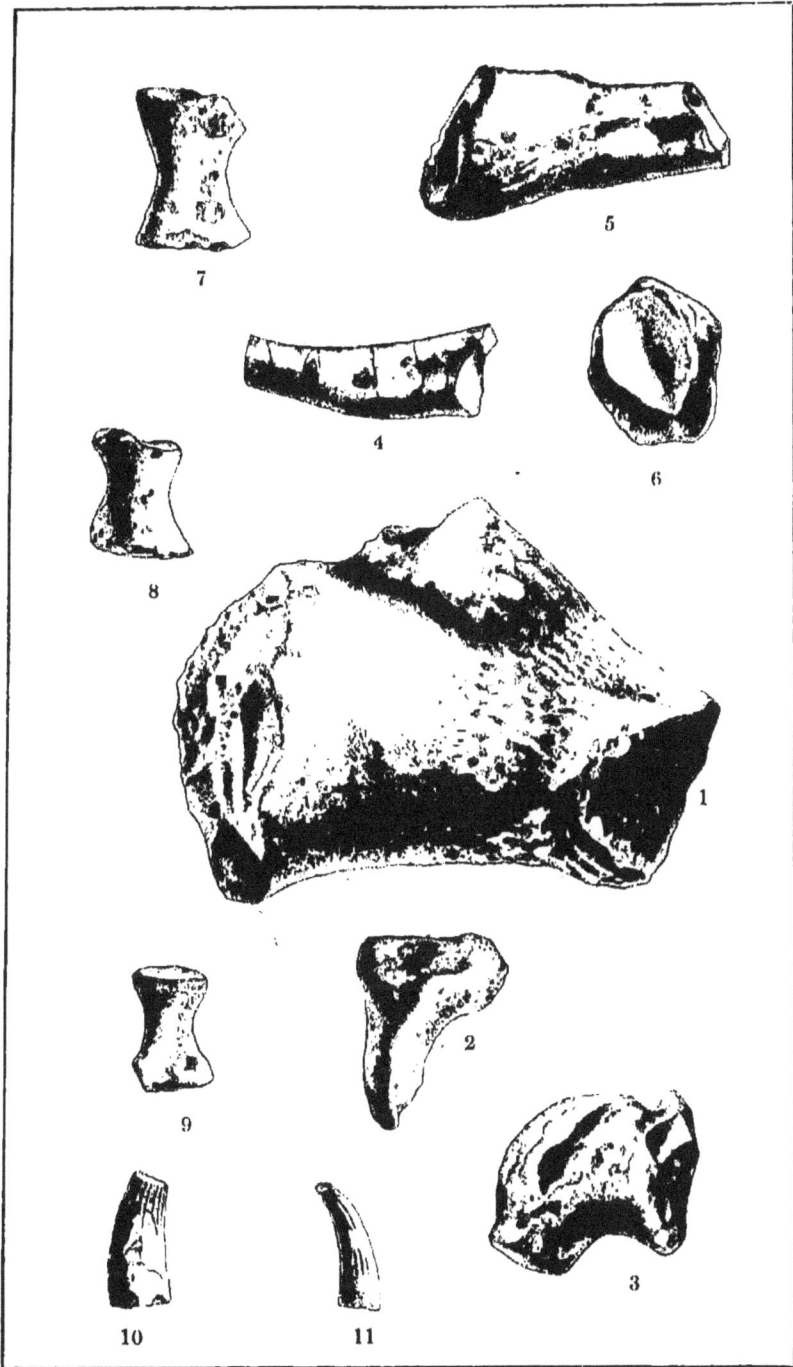

Frank R. Leggatt Del.

,г.·,, ʌ Columnopōra pauciseptata, Eth. Fil.

Fig. 1. The corallum seen from above.

" 2. The same, from one side, showing superimposed colonies.

" 3. Three calices exhibiting the septa. × 10.

" 4. A transverse section of several corallites, showing the thickened walls, primary and secondary septa, and absence of pores of communication. × 17.

" 5. A transverse section of a single corallite. × 30.

" 6. A transverse section of another corallite with the walls still more thickened, and the septa club-shaped distally. × 30.

" 7. A vertical section of several corallites, exhibiting old visceral chambers, thickened walls, and tabulæ. × 10.

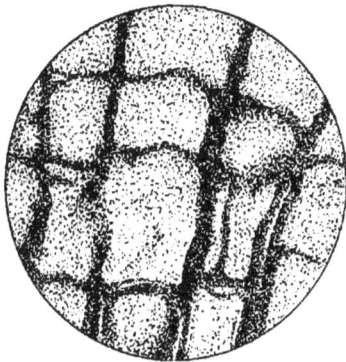

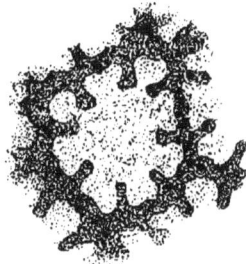

7

6

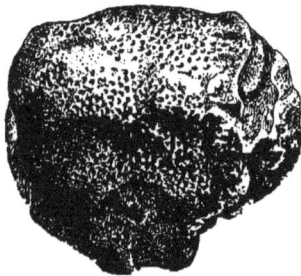

3

2

1

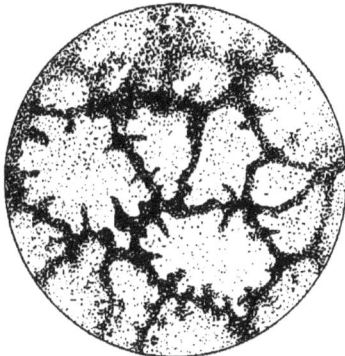

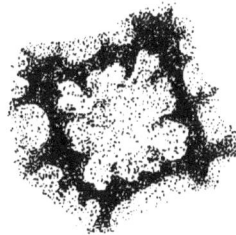

4

5

PLATE IX.

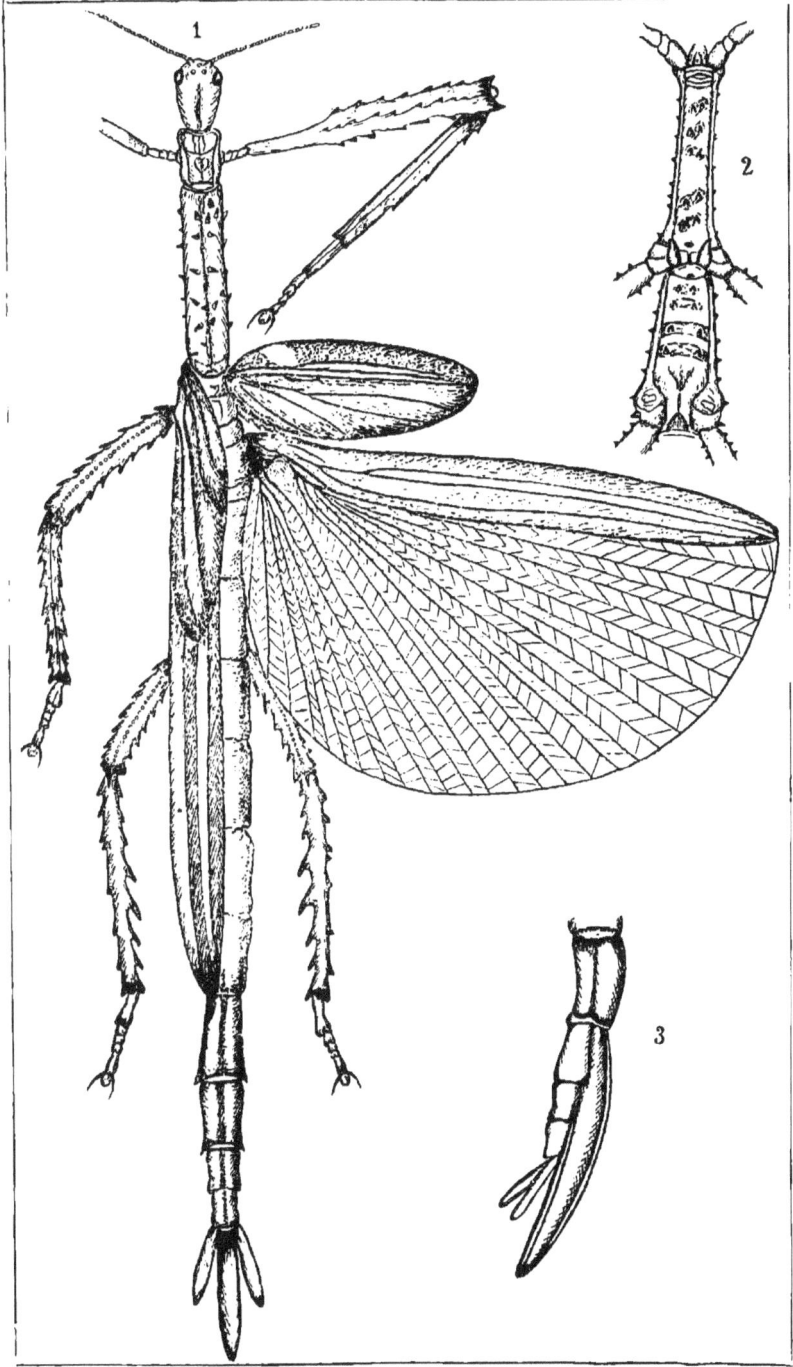

EXPLANATION OF PLATE X.

Fig. 1. *Tropidoderus decipiens*, Rainb.

,, 2. ,, ,, left tegmina, showing scheme of colouration.

,, 3. ,, ,, underside of thorax.

,, 4. ,, ,, ovipositor.

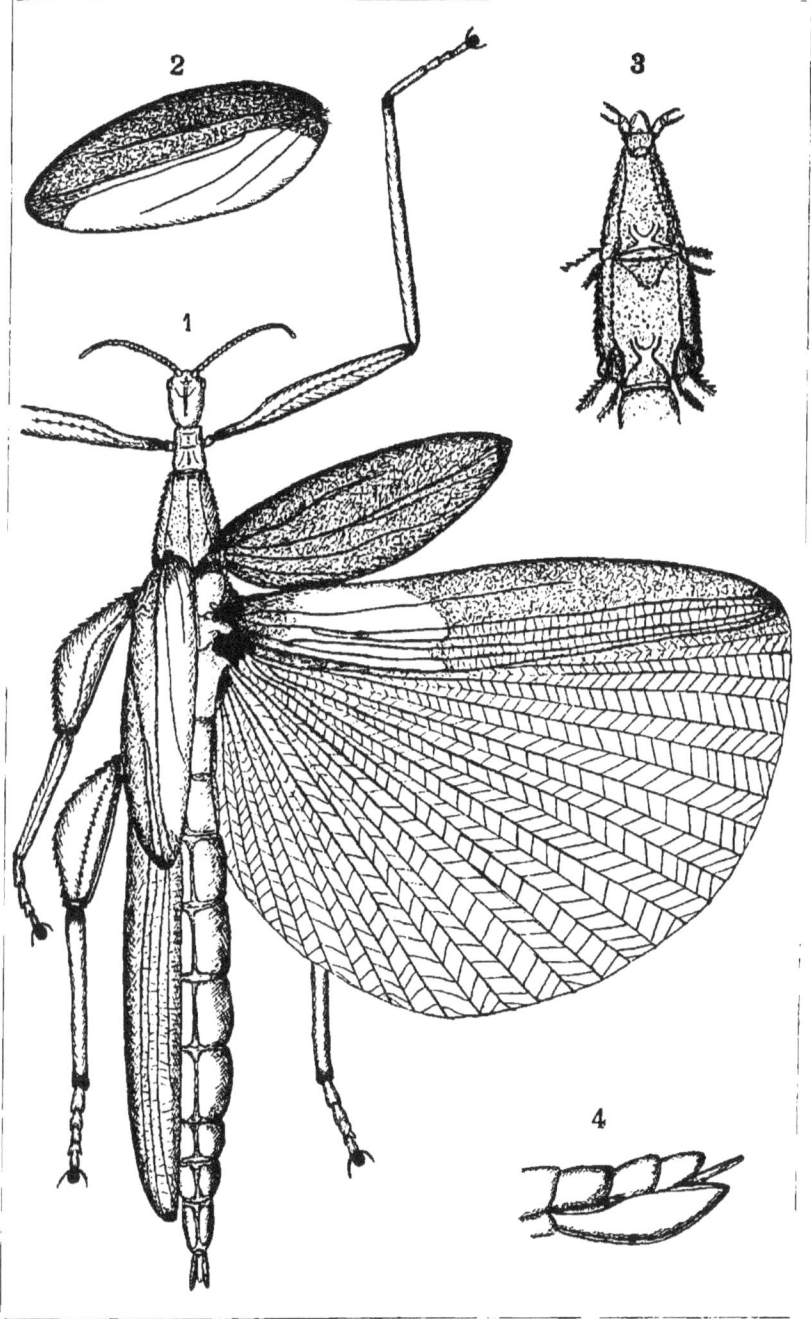

W. J. RAINBOW, del.

EXPLANATION OF PLATE XI.

Figs. 1, 2, 3. Different aspects of *Pupisoma circumlitum.*

.,   4, 5, 6. Different aspects of *Sitala ? sublimais.*

,,   7, 8, 9. Different aspects of *Endodonta waterhousiæ.*

,,   10, 11, 12. Different aspects of *Flammulina abdita.*

,,   13, 14. Jaw and part of reproductive system of *E. waterhousiæ.*

All much magnified, drawn from nature and to various scales by
C. Hedley.

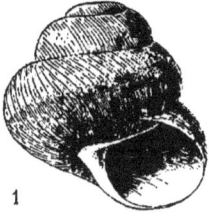

1

2

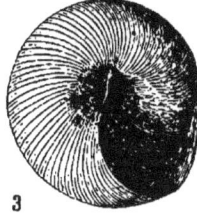

3

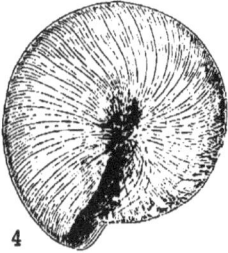

4

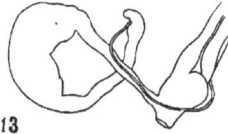

13

14

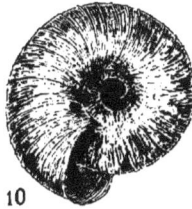

10

7

5

11

8

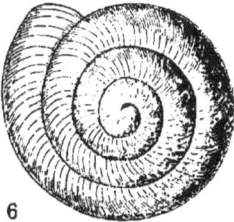

6

12

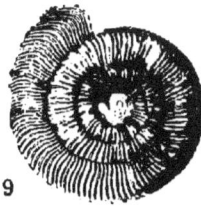

9

C. HEDLEY, del.

EXPLANATION OF PLATE XII.

Views of the Nocoleche Meteorite before cutting.

Fig. 1. Side view.

Fig. 2. Under surface.

Reproduced from photographs, about one-third natural size.

1.

2.

Etched surface of the Nocoleche Meteorite, showing Widmanstätten figures.

Reproduced from photograph, about two-thirds natural size.

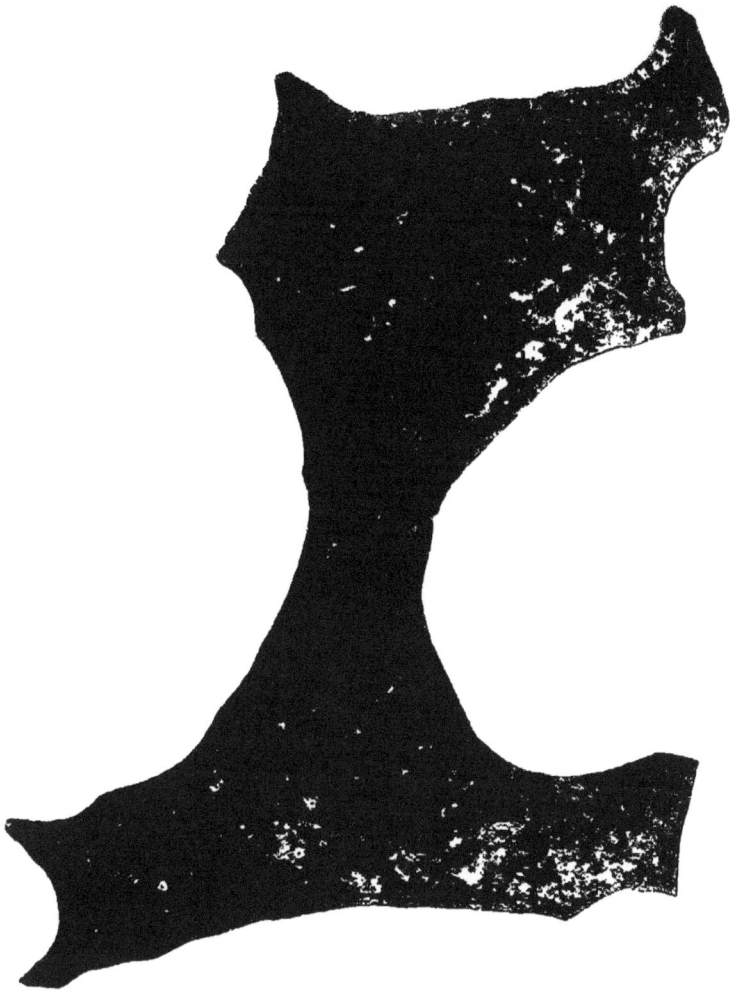

EXPLANATION OF PLATE XIV.

Same surface of the Nocoleche Meteorite as Plate XIII., but with different lighting.

Reproduced from photograph, about two-thirds natural size.

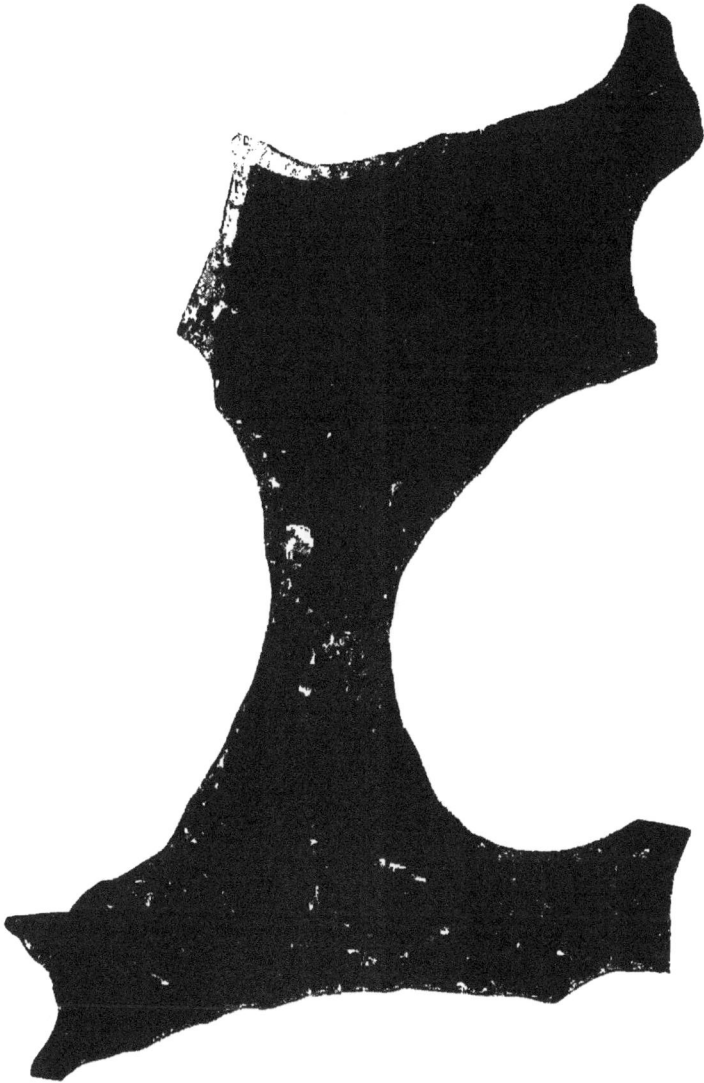

Goniostropha pritchardi, Eth. fil.

Fig. 1. Shell seen from the front, showing mouth.

,, 2. A similar example to Fig. 1.

,, 3. Shell seen from the back.

,, 4. A whorl, showing the amount of angularity, and position and nature of the band. x 2.

Mourlonia duni, Eth. fil.

,, 5. Imperfect shell, seen from the back; the sutural band is plainly distinguishable.

Helicotoma johnstoni, Eth. fil.

,, 6. Upper side, showing depressed spire.

,, 7. Under side with umbilicus.

,, 8. Back, exhibiting great width, perhaps somewhat assisted by distortion.

Trochonema ? nodosa, Eth. fil.

,, 9. Imperfect shell, seen from the back.

,, 10. The same, seen from the front.

Holopea wellingtonensis, Eth. fil.

,, 11. Natural section of the whorls, showing the form of the latter.

10

9

4

3

2

5

11

7

1

6

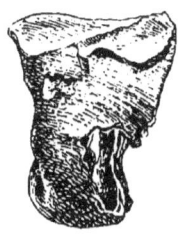

8

Frank R. Leggitt Del.

EXPLANATION OF PLATE XVI.

Gyrodoma etheridgei, Creswell.

Fig. 1. Imperfect specimen, showing a distinct and flattened band.

Mourlonia duni, Eth. fil.

„ 2. A small example with acute spire.

Helicotoma johnstoni. Eth. fil.

„ 3. Another example viewed from the upper side, exhibiting the spire.

„ 4. The under or umbilical side of the same, showing a sutral depression and telescopic umbilicus.

Trochonema etheridgei, Johnston.

„ 5. Shell seen from the front, showing mouth and keels.

„ 6. The same, seen from the back.

Holopea wellingtonensis, Eth. fil.

„ 7. A full-grown example, seen from the front.

„ 8. The same, seen from the back.

„ 9. A smaller specimen, seen from the back.

All the specimens are in the Australian Museum, except Pl. xvi., Fig. 2, which is from the Collection of the Geological Survey of N.S. Wales.

4

3

9

2

1

5

6

8

7

Frank R. Leggatt Del.

EXPLANATION OF PLATE XX.

Testudo nigrita, Dum. & Bibr.

(Gigantic Tortoise).

From a Photograph, lent by Miss Annie C. F. MacDonald.

EXPLANATION OF PLATE XIX.

Drey of Ring-tailed Opossum (*Pseudochirus peregrinus,* Bodd.)

[From a Photograph by H. Barnes, Junr., Australian Museum]

Presented by Mr. J. M. Cantle.

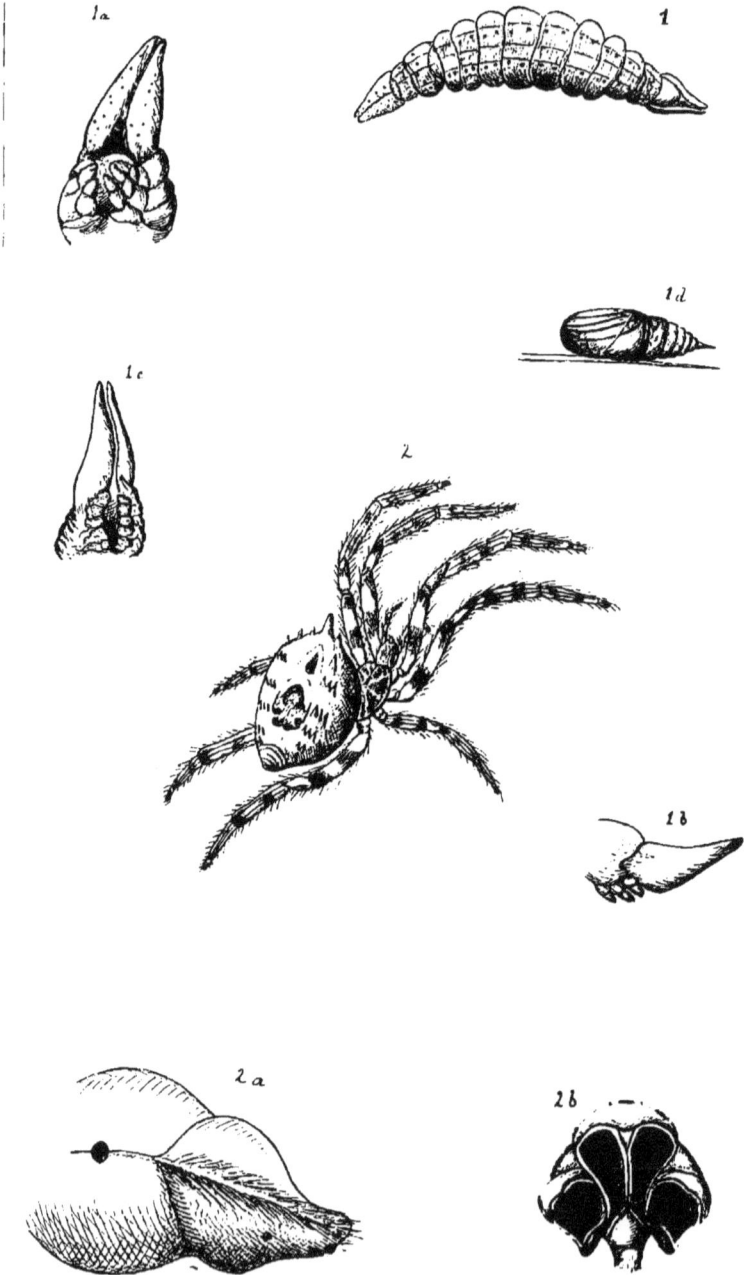

Fig. 1. Larva of *Pseudoterpna percomptaria,* Gn.
,, 1*a*. ,, ,, ,, Head, from underneath.
., 1*b*. ., ,, Head in profile.
,, 1*c*. ,. ,, ,, Tail, from underneath.
,, 1*d*. ,, ,, ,, Pupa.

,, 2. *Poltys multituberculata,* Rainbow.
,, 2*a*. ,, ,, Cephalothorax, showing arrangement of eyes as viewed from
the side.
,, 2*b*. ,, ,, Epigyne.

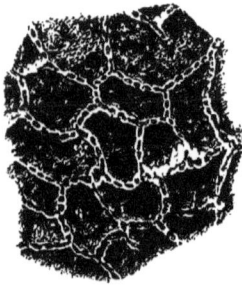

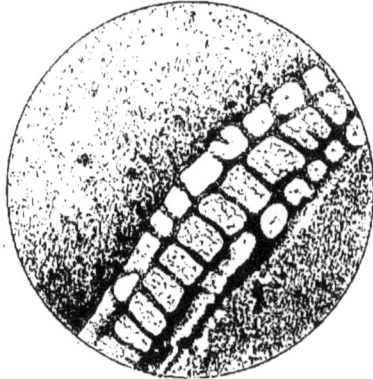

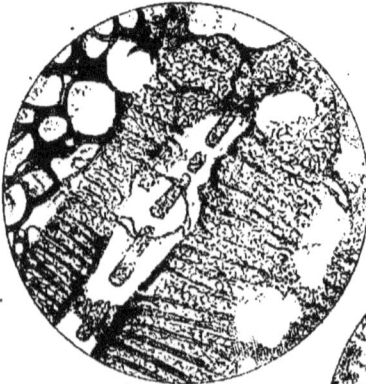

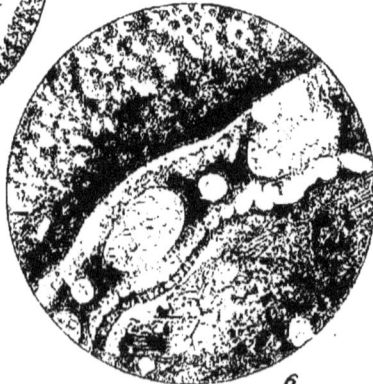

Frank R. Leggott del.

Halysites australis, Eth. fil.

Fig. 1. Portion of a corallum, seen from above, showing the form of the reticulating laminæ, and the normal corallites.

,, 2. Small portion of the same with the normal corallites and the interstitial tubes. Enlarged.

,, 3. Portion of another corallum differing from that seen in fig. 1 only in the size of the reticulating laminæ.

,, 4. The same. Enlarged.

,, 5. Portion of another corallum seen obliquely from the side, showing the lateral faces of the laminæ, and outline of the normal corallites.

,, 6. Horizontal section of part of a lamina, with two normal corallites and two interstitial tubes filled with crystalline calcite. Parts of the lamina walls are converted into chalcedony, particularly on the upper right hand; two blebs are also visible on the opposite side. × 18 (about).

,, 7 Vertical section showing two normal corallites with close horizontal tabulæ, between them an interstitial tube with distant tabulæ. The intertabular, or visceral cavities of the normal corallites are filled with crystalline calcite; the interstitial tube on the extreme right has been converted into a mass of chalcedony, the walls and tabulæ of that in the centre are in the same condition, whilst scattered on the left hand are blebs of chalcedony. × 20 (about).

,, 8. Vertical section of a normal corallite, highly altered. The tabulæ and portions of the walls remain as sclerenchyma, the other parts of the latter being converted into blebs of chalcedony. The intertabular, or visceral spaces, are filled with crystalline calcite. × 20 (about).

Testudo nigrita, Dum. & Bibr.

(Gigantic Tortoise).

Fig. 1. Male.   Fig. 2. Female.

[From Photographs taken by the Author].

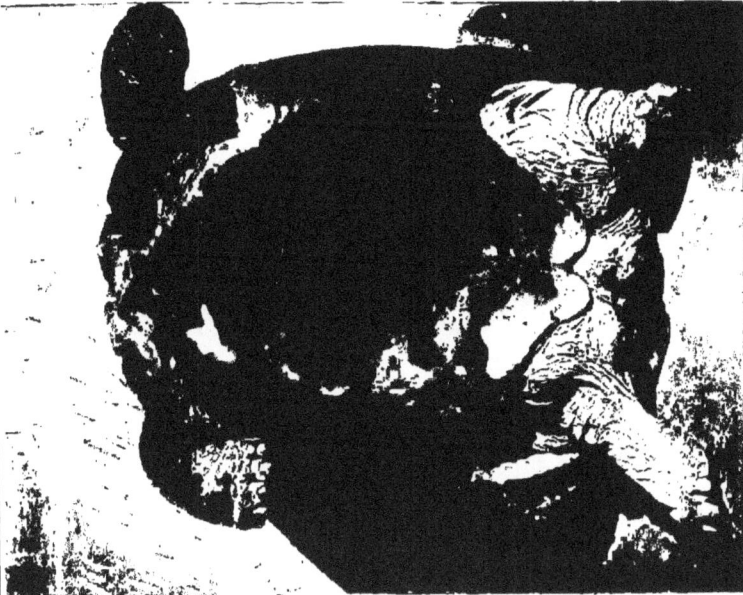

Fig. 2.

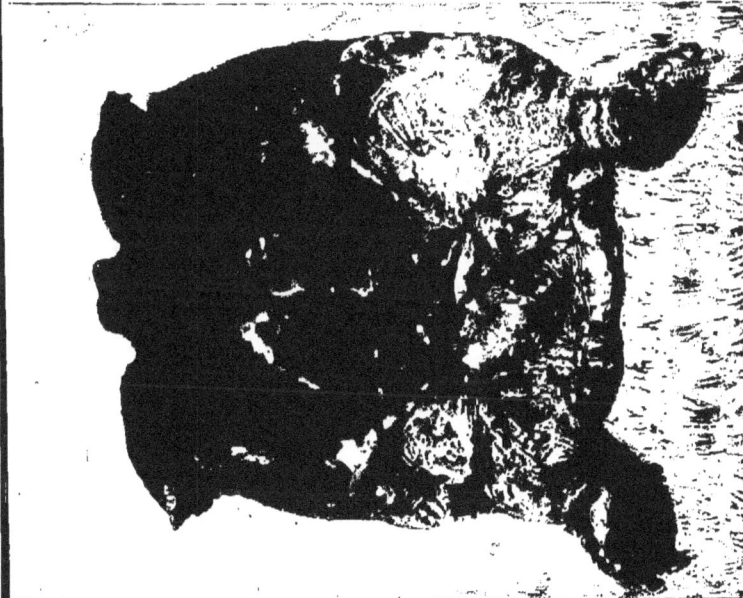

Fig. 1.

EXPLANATION OF PLATE XXII.

Testudo nigrita, Dum. & Bibr.

(Gigantic Tortoise).

Skeleton of female in the Australian Museum.

[From a Photograph by H. Barnes, Junr., Australian Museum].

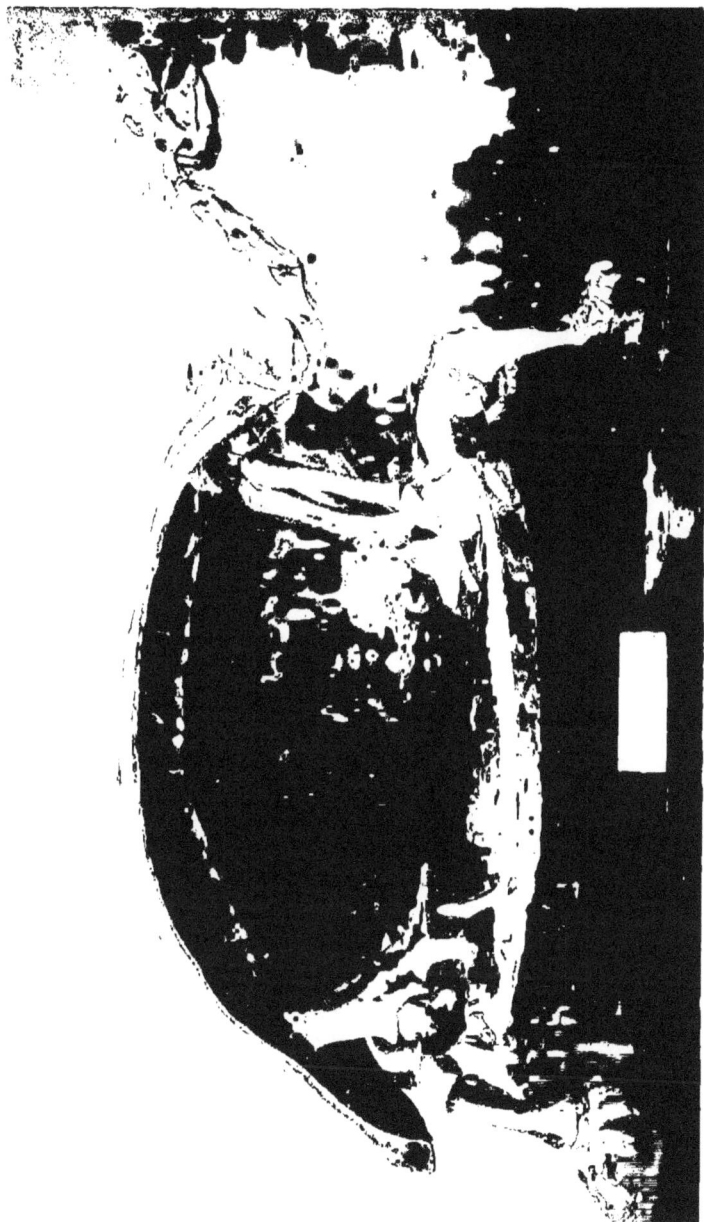

Palæopede whiteleggei, Eth. fil.

Fig. 1. Trichome (?) of moniliform cells, and heterocysts (?) ·5 mm. long.

,, 2. Four heterocysts (?) at the end of a non-segmented tube filled with similar black pulverulent matter to themselves. ·1 mm. long, diameter ·02.

,, 3. Five moniliform cells and a heterocyst (?) terminating in a clear tube. ·07 mm. long.

4. Moniliform cells enclosed in a sheath or vagina. ·08 mm. long.

Palæachlya torquis, Eth. fil.

5. Tubes filled with yellow granular matter. ·01 mm. in diameter.

[From drawings by Mr. Edgar R. Waite, Australian Museum].

2.

4.

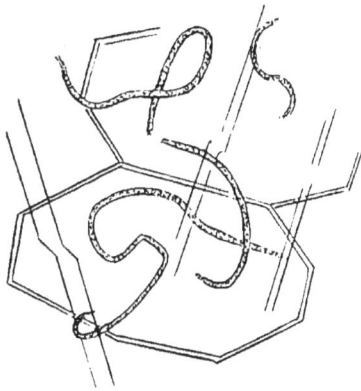

3.   1.   5.

EXPLANATION OF PLATE XXIV.

Blechnoxylon talbragarense, Eth. fil.

Fig. 1. A cluster of fronds radiating from the caudex, which is partially hollow through disintegration; seen from above; × 2½.

,, 2. A similar specimen, in which the caudex projects slightly above the level of the fronds; seen from above; × 2.

,, 3. The impressions of the upper surfaces of two more or less pyriform fronds, and portion of a third; the caudex is again partially hollowed by disintegration; × 3.

,. 4. A series of frond impressions, radiating from a caudex at three successive levels, as displayed by the shading; × 4.

,, 7. Portion of an internode seen in profile, with portions of three fronds seen from above; × 2.

,, 9. The centre of Fig. 2, showing the crushed in zones comprising the stem; a naturally weathered section; × 9.

Blechnoxylon talbrayarense, Eth. fil.

Fig. 5. An internode, partly in the round and partly as an impression, and two nodes, the latter with fronds attached in a greater or less degree. In the centre of the internode is a small protuberance that may be the base of attachment of an adventitious root. The fronds on the right have been pressed back out of the normal position, and all have suffered from the disintegration of their parts by weathering. One of those on the left has been so much decomposed that only the outline remains, with the revolute margin subdivided by the impressions of the veinules; × 3½.

Blechnoxylon talbragarense, Eth. fil.

Fig. 6. The impression of part of an internode, with a node in the round, bearing a series of leaf-scars, spirally arranged, and each pierced by a single vascular opening ; to some of the scars are attached the frond petioles. Between the fronds above are two scales similar to that seen in Fig. 5 ; × 4.

„ 8. An internode and portions of two others, more or less decorticated; two nodes are indicated by the remains of two fronds projecting from either side the stem ; × 2½.

„ 10. Section of the stem prepared for the microscope. The black centre represents either the pith, or the pith and primary wood, whichever may have existed; the second zone with radii is the secondary wood ; the third zone of varying thickness, caused probably by partial decomposition and extraneous pressure, occupies the position of the phloem in the stem of *Botrychium* ; whilst the fourth, or outer zone, represents the parenchyma of the latter. The endo- and exodermal layers are represented by the two outermost irregular rings; highly enlarged.

„ 11. A portion of the two innermost zones of Fig. 10, the dark centre being the pith, or pith and primary wood, as the case may be, and the outer radial cellular portion the secondary wood ; very highly enlarged.

„ 17. The revolute end ot the right-hand side of Fig. 16 (Pl. xxvii.) showing the supposed pedicels, one of them supporting a round body that may be a sporangium ; very highly enlarged.

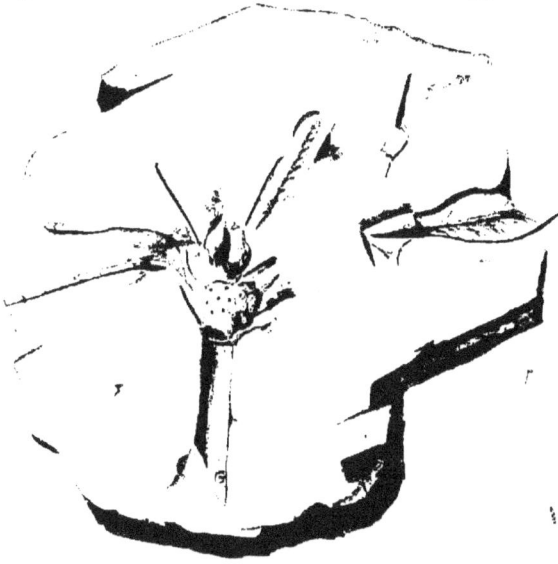

6

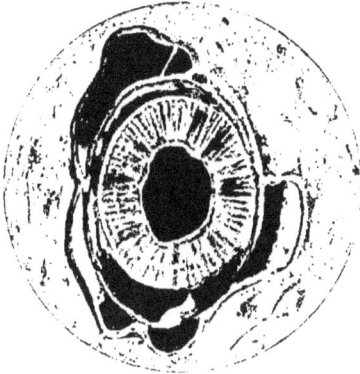

10

8

11

17

Blechnoxylon talbragarense, Eth. fil.

Fig. 12. Longitudinal section of portion of a caudex, prepared for the microscope. The centre represents either the pith, or the pith and primary wood, which ever may have existed; the second zone with longitudinal lines is the secondary wood. At the sides are probably seen the bases of petioles; highly enlarged.

,, 13. A root and rootlets, but whether or no of this organism it is impossible to say; highly enlarged.

,, 14. Horizontal section of a frond, prepared for the microscope, showing on the right-hand side epidermal tissue, and on the left dark brown patches between the veinules; highly enlarged.

15. Longitudinal section of a frond, prepared for the microscope, showing the upper and lower surfaces clothed with setiform hairs, internal "tissue pillars," alternating with clear vacuities, etc.; highly enlarged.

,, 16. Cross section of a frond prepared for the microscope, showing the form of the frond, its revolute margins; and on the right-hand side, filaments that may be pedicels for the support of sporangia; highly enlarged.

13.   14

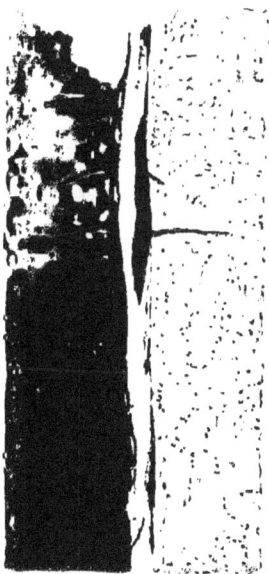

16

15   17

EXPLANATION OF PLATE XXVIII.

Figs. 1, 2, 3. Various aspects of *Endodonta aculeata*, Hedley.

,,   4, 5, 6. Various aspects of *Endodonta norfolkensis*, Hedley.

,,   7. Jaw of *Papuina hindei*, Cox.

,,   8. Genitalia of the same.

,,   9. Portion of radula of the same.

,,   10, 11. Two aspects of *Papuina mayana*, Hedley.

,,   12, 13. Two aspects of *Dendrotrochus mentum*, Hedley.

,,   14. *Tornatellina wakefieldæ*, Cox.

Figs. 10 and 11, natural size; the remainder enlarged to various scales.

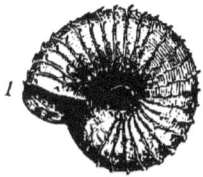
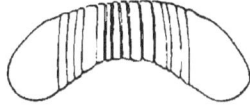

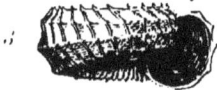

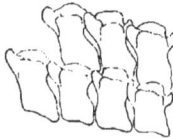

Scyllarus sculptus, Latr.

Male, two-thirds the natural size, seen from the dorsal side.

EXPLANATION OF PLATE XXX.

Fig. 1. *Misumena tristania*, ♀, Rainbow.
Fig. 1*a*.   ,,   ,,   Epigyne.
Fig. 2. *Saccodomus formivorus*, ♀, Rainbow.
Fig. 2*a*.   ,,   ,,   Cephalothorax, profile.
Fig. 2*b*.   ,,   ,,   Abdomen, profile, normal.
Fig. 2*c*.   ,,   ,,   Abdomen, profile, gravid.
Fig. 2*d*.   ,,   ,,   Epigyne.
Fig. 2*e*.   ,,   ,,   Eyes.

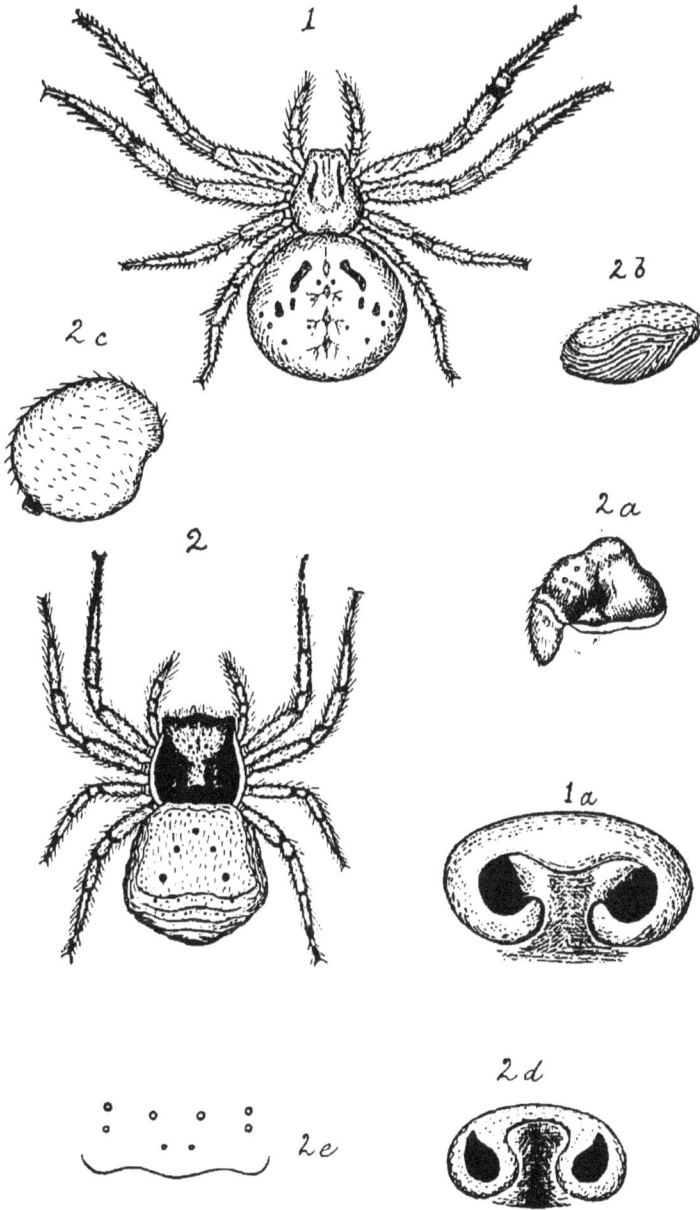

Stutchburia costata, Morris, sp.

Fig. 1. Internal cast of a left valve.

Stutchburia compressa, Morris, sp.

„ 2. Internal cast of a left valve. A = crest representing the ligamental (?) pit in front of the umbo.

Stutchburia obliqua, Eth. fil.

„ 3. Internal cast of a right valve.

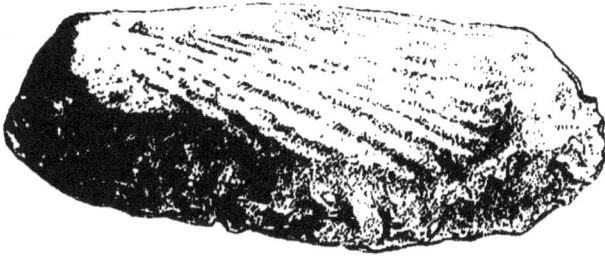

1.

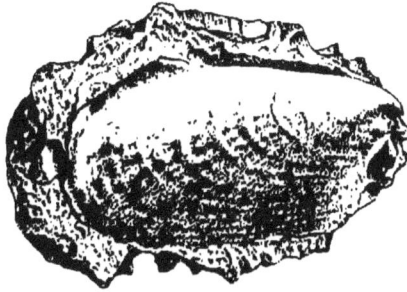

3.

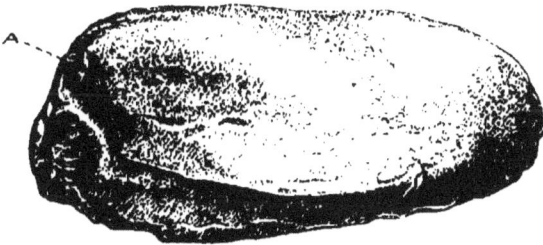

2.

Limoptera ? permo-carbonifera, Eth. fil.

Fig. 1. Left valve with indistinct radii.

„ 2. Right valve with ligamental pit under the left umbo and nodes representing muscular pits in the umbonal cavity of the right valve.

Stutchburia farleyensis, Eth. fil.

„ 3. Internal cast of a right valve.

„ 4. Hinge line of conjoined valves, internal cast.

„ 5. Internal cast of anterior end of conjoined valves.

„ 6. The same of another example.

5.   6

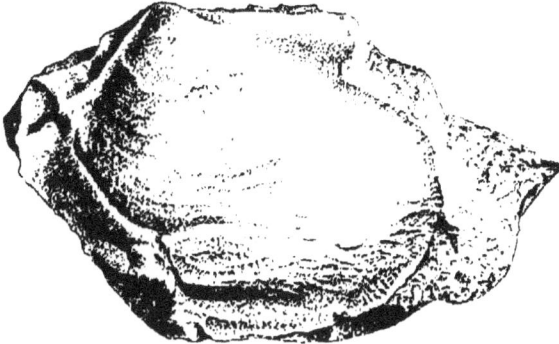

1.

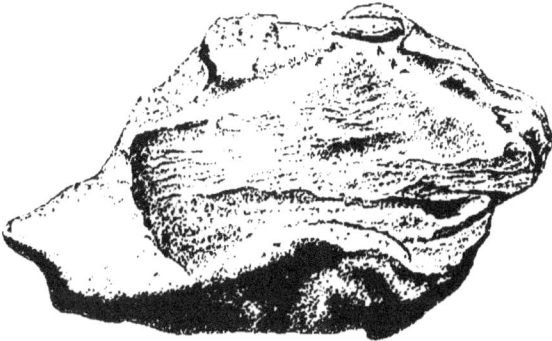

2.

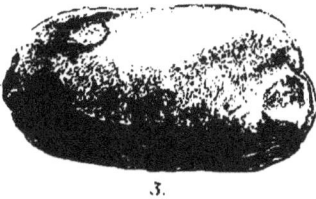

3.   4.

Stutchburia compressa, Morris, sp.

Fig. 1.   Internal cast of conjoined valves showing the hinge line

Pleurophorus gregarius, Eth. fil.

,,   2.   Internal cast of a right valve.

,,   3.   Internal cast of the anterior end of the conjoined valves, showing the impressions of the cardinal teeth.

,,   4.   Internal cast of a left valve, with the impressions of the lateral teeth.

,,   5.   Similar to Fig. 3.

Mytilops ? ravensfieldensis, Eth. fil.

,,   6.   Internal cast of a left valve.

,,   7.   Anterior end showing a cardinal fold.   × 3.

4.

3

2.

/

5.

/.

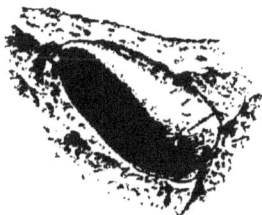

6

EXPLANATION OF PLATE XXXIV.

Amphiprion latezonatus, Waite.

Natural size.

[Reproduced from a drawing by the author.]

EXPLANATION OF PLATE XXXV.

Holacanthus conspicillatus, Waite.
Three-fourths natural size.

[Reproduced from a drawing by the author.]

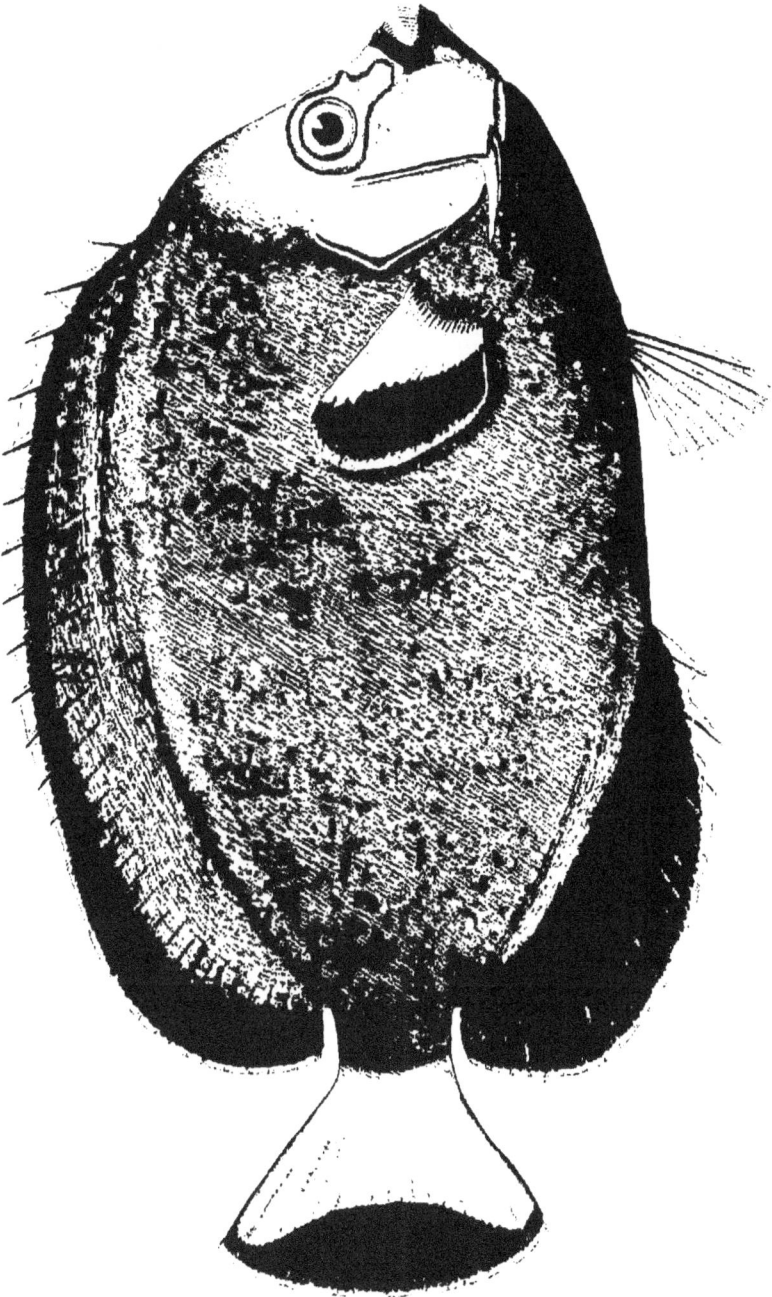

Holacanthus semicinctus, Waite.

Four-fifths natural size.

[Reproduced from a drawing by the author.]

EXPLANATION OF PLATE XXXVII.

Hoplegnathus woodwardi, Waite.
One-half natural size

[Reproduced from a drawing by the author.]

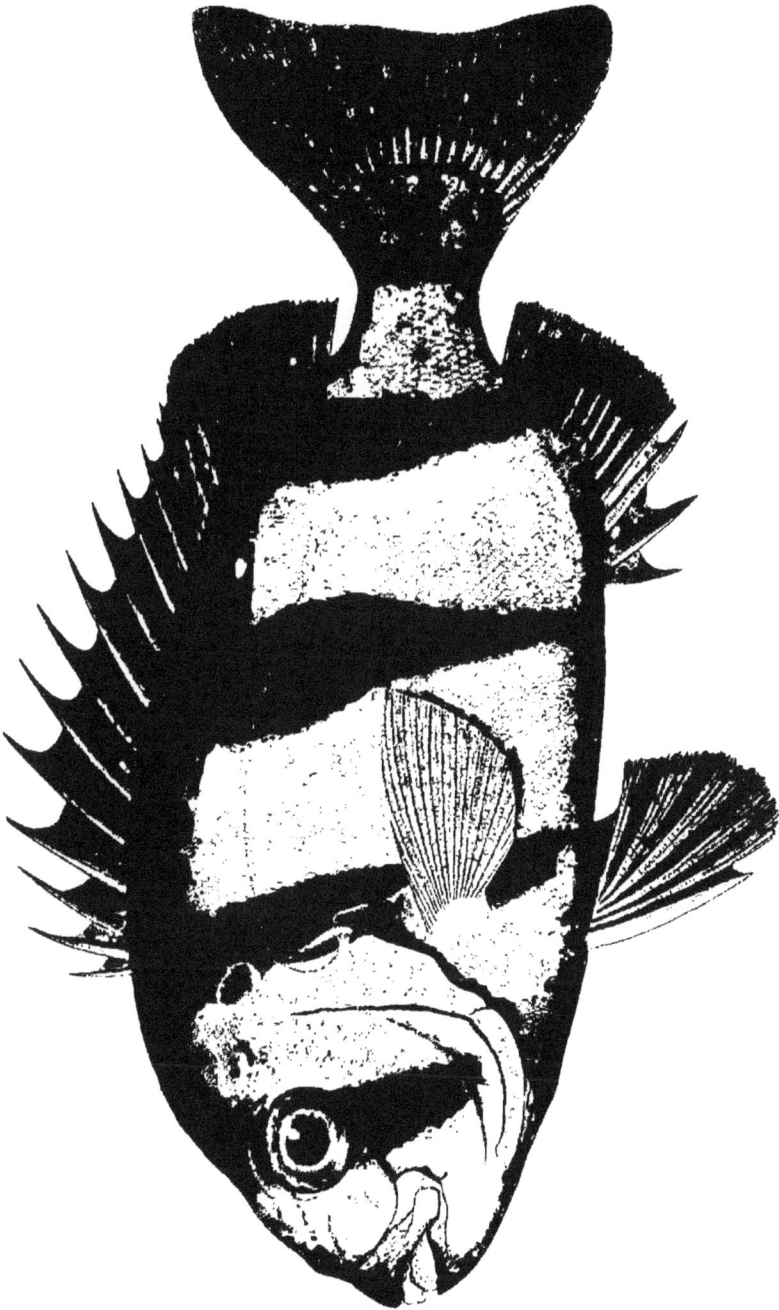

INDEX.